Ethics Lost in Modernity

VERITAS
Series Introduction

"... the truth will set you free" (John 8:32)

In much contemporary discourse, Pilate's question has been taken to mark the absolute boundary of human thought. Beyond this boundary, it is often suggested, is an intellectual hinterland into which we must not venture. This terrain is an agnosticism of thought: because truth cannot be possessed, it must not be spoken. Thus, it is argued that the defenders of "truth" in our day are often traffickers in ideology, merchants of counterfeits, or anti-liberal. They are, because it is somewhat taken for granted that Nietzsche's word is final: truth is the domain of tyranny.

Is this indeed the case, or might another vision of truth offer itself? The ancient Greeks named the love of wisdom as *philia*, or friendship. The one who would become wise, they argued, would be a "friend of truth." For both philosophy and theology might be conceived as schools in the friendship of truth, as a kind of relation. For like friendship, truth is as much discovered as it is made. If truth is then so elusive, if its domain is *terra incognita*, perhaps this is because it arrives to us—unannounced—as gift, as a person, and not some thing.

The aim of the Veritas book series is to publish incisive and original current scholarly work that inhabits "the between" and "the beyond" of theology and philosophy. These volumes will all share a common aspiration to transcend the institutional divorce in which these two disciplines often find themselves, and to engage questions of pressing concern to both philosophers and theologians in such a way as to reinvigorate both disciplines with a kind of interdisciplinary desire, often so absent in contemporary academe. In a word, these volumes represent collective efforts in the befriending of truth, doing so beyond the simulacra of pretend tolerance, the violent, yet insipid reasoning of liberalism that asks with Pilate, "What is truth?"—expecting a consensus of non-commitment; one that encourages the commodification of the mind, now sedated by the civil service of career, ministered by the frightened patrons of position.

The series will therefore consist of two "wings": (1) original monographs; and (2) essay collections on a range of topics in theology and philosophy. The latter will principally be the products of the annual conferences of the Centre of Theology and Philosophy (www.theologyphilosophycentre.co.uk).

Conor Cunningham and Eric Austin Lee, *Series editors*

Not available from Cascade Books

Deane-Peter Baker	*Tayloring Reformed Epistemology: The Challenge to Christian Belief.* Volume 1
P. Candler & C. Cunningham (eds.)	*Belief and Metaphysics.* Volume 2
P. Candler & C. Cunningham (eds.)	*Transcendence and Phenomenology*
Marcus Pound	*Theology, Psychoanalysis, and Trauma.* Volume 4
Espen Dahl	*Phenomenology and the Holy.* Volume 5
C. Cunningham et al. (eds.)	*Grandeur of Reason: Religion, Tradition, and Universalism.* Volume 6
A. Pabst & A. Paddison (eds.)	*The Pope and Jesus of Nazareth: Christ, Scripture, and the Church.* Volume 7
J. P. Moreland	*Recalcitrant Imago Dei: Human Persons and the Failure of Naturalism.* Volume 8

Available from Cascade Books

[Nathan Kerr	*Christ, History, and Apocalyptic: The Politics of Christian Mission.* Volume 3][1]
Anthony D. Baker	*Diagonal Advance: Perfection in Christian Theology.* Volume 9
D. C. Schindler	*The Perfection of Freedom: Schiller, Schelling, and Hegel between the Ancients and the Moderns.* Volume 10
Rustin Brian	*Covering Up Luther: How Barth's Christology Challenged the* Deus Absconditus *that Haunts Modernity.* Volume 11
Timothy Stanley	*Protestant Metaphysics After Karl Barth and Martin Heidegger.* Volume 12
Christopher Ben Simpson	*The Truth Is the Way: Kierkegaard's* Theologia Viatorum. Volume 13
Richard H. Bell	*Wagner's Parsifal: An Appreciation in the Light of His Theological Journey.* Volume 14
Antonio Lopez	*Gift and the Unity of Being.* Volume 15
Toyohiko Kagawa	*Cosmic Purpose.* Translated and introduced by Thomas John Hastings. Volume 16
Nigel Zimmerman	*Facing the Other: John Paul II, Levinas, and the Body.* Volume 17

1. Note: Nathan Kerr, *Christ, History, and Apocalyptic*, although volume 3 of the original SCM Veritas series, is available from Cascade as part of the Theopolitical Visions series.

Conor Sweeney	*Sacramental Presence after Heidegger: Onto-theology, Sacraments, and the Mother's Smile.* Volume 18
John Behr et al. (eds.)	*The Role of Death in Life: A Multidisciplinary Examination of the Relation between Life and Death.* Volume 19
Eric Austin Lee et al. (eds.)	*The Resounding Soul: Reflection on the Metaphysics and Vivacity of the Human Person.* Volume 20
Orion Edgar	*Things Seen and Unseen: The Logic of Incarnation in Merleau-Ponty's Metaphysics of Flesh.* Volume 21
Duncan B. Reyburn	*Seeing Things as They Are: G. K. Chesterton and the Drama of Meaning.* Volume 22
Lyndon Shakespeare	*Being the Body of Christ in the Age of Management.* Volume 23
Michael V. Di Fuccia	*Owen Barfield: Philosophy, Poetry, and Theology.* Volume 24
John McNerney	*Wealth of Persons: Economics with a Human Face.* Volume 25
Norm Klassen	*The Fellowship of the Beatific Vision: Chaucer on Overcoming Tyranny and Becoming Ourselves.* Volume 26
Donald Wallenfang	*Human and Divine Being: A Study of the Theological Anthropology of Edith Stein.* Volume 27
Sotiris Mitralexis	*Ever-Moving Repose: A Contemporary Reading of Maximus the Confessor's Theory of Time.* Volume 24
Sotiris Mitralexis et al. (eds.)	*Maximus the Confessor as a European Philosopher.* Volume 28
Kevin Corrigan	*Love, Friendship, Beauty, and the Good: Plato, Aristotle, and the Later Tradition.* Volume 29
Andrew Brower Latz	*The Social Philosophy of Gillian Rose.* Volume 30
D. C. Schindler	*Love and the Postmodern Predicament: Rediscovering the Real in Beauty, Goodness, and Truth.* Volume 31
Stephen Kampowski	*Embracing Our Finitude: Exercises in a Christian Anthropology between Dependence and Gratitude.* Volume 32
William Desmond	*The Gift of Beauty and the Passion of Being: On the Threshold between the Aesthetic and the Religious.* Volume 33
Charles Péguy	*Notes on Bergson and Descartes.* Volume 34
David Alcalde	*Cosmology Without God?: The Problematic Theology Inherent in Modern Cosmology.* Volume 35
Benson P. Fraser	*Hide and Seek: The Sacred Art of Indirect Communication.* Volume 36

Ethics Lost in Modernity

Reflections on Wittgenstein and Bioethics

MATTHEW VEST

Foreword by
JEFFREY P. BISHOP

CASCADE *Books* • Eugene, Oregon

ETHICS LOST IN MODERNITY
Reflections on Wittgenstein and Bioethics

Veritas 39

Cascade Books
An Imprint of Wipf and Stock Publishers
199 W. 8th Ave., Suite 3
Eugene, OR 97401

www.wipfandstock.com

PAPERBACK ISBN: 978-1-6667-4718-8
HARDCOVER ISBN: 978-1-6667-4719-5
EBOOK ISBN: 978-1-6667-4720-1

Cataloguing-in-Publication data:

Names: Vest, Matthew, author. | Bishop, Jeffrey P., 1967–, foreword.

Title: Ethics lost in modernity : reflections on Wittgenstein and bioethics / by Matthew Vest.

Description: Eugene, OR : Cascade Books, 2023 | Series: Veritas 39 | Includes bibliographical references and index.

Identifiers: ISBN 978-1-6667-4718-8 (paperback) | ISBN 978-1-6667-4719-5 (hardcover) | ISBN 978-1-6667-4720-1 (ebook)

Subjects: LCSH: Wittgenstein, Ludwig, 1889–1951. | Bioethics—Philosophy.

Classification: R724 .V47 2023 (print) | R724 .V47 (ebook)

07/17/23

Dedicated to my father, Dr. Richard Norris Vest II
(1949–2021)

Just because our moral philosophers knew the facts of morality only very approximately in arbitrary extracts or in accidental epitomes—for example, as the morality of their environment, their class, their church, the spirit of their time, their climate and part of the world—just because they were poorly informed and not even very curious about different peoples, times, and past ages—they never laid eyes on the real problems of morality; for these emerge only when we compare many moralities. In all 'science of morals' so far one thing was lacking, strange as it may sound: the problem of morality itself; what was lacking was any suspicion that there was something problematic here.

—FRIEDRICH NIETZSCHE, *BEYOND GOOD AND EVIL*

There is not a single philosophical method, though there are indeed methods, like different therapies, as it were.

—WITTGENSTEIN, *PI*, 133

I think I summed up my attitude to philosophy when I said: philosophy ought really to be written only as a poetic composition [Philosophie dürfte man eigentlich nur DICHTEN]. It must, as it seems to me, be possible to gather from this how far my thinking belongs to the present, future or past. For I was thereby also revealing myself as someone who cannot quite do what he would like to be able to do.

—WITTGENSTEIN, *CV*, 2

Contents

Foreword

Jeffrey P. Bishop

"The owl of Minerva begins its flight only with the onset of dusk."[1] G. W. F. Hegel's famous quip points to a perennial problem of all human endeavors. Philosophy arrives at the close of the day, after the day's work is completed. Wisdom is the thing that you get just after you need it. Philosophy is Monday-morning quarterbacking. Philosophy is always a day late and a dollar short of the reality it hopes to know. Philosophy is only possible at the end of an era. In a way, the love of wisdom, *philosophia,* is a love that pines for that which is fading into darkness. Wisdom arrives too late for bioethics as well.

In this volume, *Ethics Lost in Modernity*, Matthew Vest gives us the first sustained and thorough Wittgensteinian critique on the limits of the language games played in medicine, medical science, and bioethics. Bioethics is a self-referential language game. Yet Vest is not content to point to the limits of medicine and bioethics in the way that Wittgenstein drew attention to the limits of scientific and philosophical language. Like Wittgenstein, Vest shows that dusk need not be darkness, but rather a kind of apophatic revealing that is possible in the remains of the day, and even into the night. Vest points to the mystical heart of medicine, healthcare—even medical science—and especially ethics, asking us not to attend to the language but to the mystery of being that dances behind, and that even animates language.

1. Hegel, *Elements of the Philosophy*, 23.

* * * * *

The story of bioethics goes something like this. Medicine, playing out the language games of the Enlightenment Project to their logical conclusion, became intoxicated with its successes, small though they were. With a Whiggish historical inevitability, medical technoscience believed it was poised to arrive manifestly at humankind's destiny.

Meanwhile, no one was keeping count of the failures that had built up behind such successes. By failures, I do not mean the lack of successes, but the subtle failures. Yes, we kept people alive longer on the machines of technoscience. But then we found that a subset of patients were living lives worse than death, on those machines. Yes, we found a technology that worked for dialysis. But then we didn't have enough dialysis machines to go around, so who gets those machines? Yes, the novel technologies for transplantation arrive, but now we need organs. But then, the people that had been living on machines, who no one thought of as dead, suddenly could be declared dead, providing a source of organs. Yes, novel scientific insights helped propel medical successes. But then, there are the untold thousands upon thousands of experimental subjects—both the human subjects (Black and Jewish persons, sex-workers, orphans) and the untold number of nonhuman, animal subjects—that have been sacrificed on the altar of technological success. While the story of medicine's failures is an old story, it is one still lived out in medical technoscience. Drunk off the elixir of its manifest destiny, medical technoscience continues seeking the elixir of life, and forgets to ask if it is all worth it, or if it instead demeans human life.

However, reality always exceeds our theories of it. Or, as Theodor Adorno might put it: "objects do not go into their concepts without leaving a remainder . . . "[2] That is true not only of our theories of biology, physiology, genetics, or neuroscience, all of which must reduce life to physiochemical processes of the body; but it is also true of our theories about ethics—deontology or utilitarianism, or hedonism, or consequentialism—which must reduce morality to imperatives or to a moral calculus. In fact, in the streams of ethics that flowed out of Enlightenment theorizing—in the wake of which we are still thrown about—even the ethics of virtue became a theory of right, instead of a practice oriented toward the good of life. Virtue theory is, after all, an oxymoron; it must have a good that escapes the limits of language, or it demeans the very thing it hopes to save. Or at least that is Wittgenstein's point.

2. Adorno, *Negative Dialectics*, 5.

With the rise of medical science and technology, the humanity of the human—the human search for meaning even in the direst of straights of human frailty—became occluded in the drive to keep the human alive with the fruits of technoscience. It was the theologians who set out to recover what had fallen into the dusk of obscurity after the bright light of the Enlightenment day. They hoped to draw the attention away from the medical technoscience to the human, whose being and whose ultimate concerns transcended even life itself. They pointed out that the human being is not reducible to the idealist tendencies in all theorizing, even while such theorizing gave lip-service to a realist font of its knowledge.

To shed light on what was fading into the dusk, theologians brought an older language to the table giving credence to the idea that the whole of being human is greater than the sum of its parts. A truth had been forgotten in the midst of the reductive nature of physiology, genetics, psychology, or neuroscience. This older language was so embedded in prior cultures, it never had to be explicit. Its resonances were part of the forms of living, the very fabric of lives, reminiscent of a time before technoscientific enlightenment. Words like "*imago Dei*" or phrases like "life is gift" made sense to a larger segment of the population. The language of "covenant" would be stacked up against the language of "contract" when explaining the relationship of healer to those who needed healing, and people had a vague notion of what was meant. They spoke to something ineffable about the human.

The use of theologically embedded language seemed to capture some imaginations, resurrecting for contemporary consumption the proverbial wisdom from the ages. Yet, in time, all language falls short of its objects and the intentions of language-users, and the older language games did not always sit well with those who think of themselves as doing the manifestly good work of saving humankind of its frailties.

The promise of technoscientific success to cure the blind, to make the lame walk, to heal the sick, and yea, even to raise the dead, was grounded in scientific theories—spawn of the Enlightenment. Science became the language of nature, a true language read off the face of nature itself. Science gives us the truth—"science is real," a political assertion, as much as it is a metaphysical and epistemological assertion; one need only adhere to the logic of the natural grammar deciphered by the scientist. The older language of "God," or "grace," or "gift," or "covenant," while lingering in the broader culture and resonating with some remnant of older webs of belief, was poorly fitted to the language of science. And with its Whiggish optimism, science does not want to get bogged down with arcane baggage.

Many of those theologians began the work of extricating the meanings of these arcane languages from their archaic webs of belief, translating these languages into idioms of the scientific day. Some quickly realized that the work of translation left behind a remainder, something that is ineffable—the human person made in the image and likeness of the Holy One, for example—exceeding the limits of all language, naming untranslatable realities. Some theologians, like Joseph Fletcher, abandoned entirely the archaic and mystical language games, entering fully into the language game of medical science. And to make matters worse, theologians were terrible at these translations, because the mystery of being remains, even in the archaic language games of theology.

Philosophers soon arrived, believing themselves able to get to the metaphysical heart. They claimed a neutral position to rescue the metaphysical truths—the true concepts—from the archaic, imprecise, and metaphorical theological language. Like Bertrand Russell, they believed that the language of analytic philosophy must become the foundation for the language of science. Philosophy itself must become like science.

Still other more cosmopolitan theologians and philosophers arrived on the scene. Rather than merely translating archaic metaphysical language, they promised to find the common moral language, a language fit for the technoscientific future. One theologian (James Childress, a deontologist) and one philosopher (Tom Beauchamp, a utilitarian) found a way to decode the thick metaphysical moral languages of religion, distilling these religious concepts to words better fit for service to medical science.

The Kantian concept of "respect for persons"—already an abstracted concept of a religion within the limits of reason alone—had to be further gutted of its metaphysically thick content, and narrowed further to autonomy. Justice, no longer filled out and content-full, no longer rolling down like the rivers, would become egalitarian fairness. Each abstracted concept of a prior religious origin took up its place, representing the tension between the one and the many—autonomy, on one hand, and justice, on the other. These two abstracted principles would now be able to create the deontological boundary within which the utilitarian calculus of beneficence and nonmaleficence could sum up the way forward. Medicine could continue unabated, while announcing that the wisdom of Minerva had set out the lineaments for further development in science.

What neither the theologians nor the philosophers noticed is that Minerva's owl had taken flight too late. If one has to remind medicine or medical science that the patient is a person, is it not too late in the day to set out the definition of a person? If one must define dignity in order to prevent someone from killing a patient, has the game not already been lost? More importantly, Vest asks the question: can any language disrupt the powerful language game of technoscientific morality? Or does language itself open to the darkness? Or is it that what appears as darkness to the "Enlightenment" is really the light of the apophatic?

* * * * *

Wittgenstein's task was to point out that language is not what the logical positivists say it is. There is no one-to-one correspondence between a word and a thing, a word-thing, verifiable in sense impression, strictly delimited, and in lockstep with logic. Russell's task then was to dethrone metaphysics and to make philosophy strictly speaking physics. Wittgenstein called foul: language understood thusly cannot get to the ethical heart of the matter. Language, as it is imagined by the Enlightenment and post-Enlightenment philosophy, whether in the German idealist strand of philosophy or the logical positivist and analytic strand, cannot get to being, nor can it get to the good.

Certainly, it is true that, as Vest points out, language is the source of technoscientific idols, to which it sacrifices humans and other animals for some technoscientific god-like status to control living and dying. Or think of the fact that human meaning, a life well-lived, is sacrificed to technoscientific control of the dying in the ICU. Certainly, language can enact, for a time and within a set of delimited conditions, what technoscience wants. Yet language is actually more than the logical positivist imagines. Wittgenstein and Vest both note that language can do the work that technoscience sets out for it to do, but the kind of seeing that this language makes possible also blinds us.

But all is not lost. There is more to language than meets the technoscientific/analytic philosophical eye. Language is not merely a link that stands between subject and object, or a lens through which a subject peers at its object. It is not merely the medium between subjects and objects, between observer and observed, between agent and the object of its intention. Language is inside us, in that we each have a particular language, learned from a time immemorial—in both the historical sense but also in a psychological sense in that there is not a time I can remember in which I did not have language. Our concepts are put into language, even if we try to formalize it in the austerity of logic. Language is also outside us in that we are thrown into a language—constitutive of a culture and also at the same time constituted by a culture—not of our own choosing.

Words are at play between the one and the many, the particular and the universal. A word connects the subject to the object; it has something of both in it. Words shed light on what it is possible to see. Words capture something of the object for us to bring it into communion with our own being. Words beget words. In other words, we always put words into other words. Word calls to us to make more words.

Thus, language is active; it is sometimes under the subject's control, and sometimes not. Language sometimes reveals the truth about an object or about the good, and sometimes obscures it. Beings and the good are also active, in that sometimes beings lend themselves to human language, and sometimes they do not. Objects in fact do not go easily into our concepts of them, and yet, there is a way that even the unknown makes its way into language. The good does not always, nor easily, make its way into the moral theories—deontological, utilitarian, consequentialist—or proceduralisms of bioethics.

Heidegger, late in his life, saw the vocation of the poet to be one that produced novel insight about being, through the language that poets deployed. Being worms its way into language, but always exceeds it, through the poet. Language conjures; it calls being into relief. Or, as Heidegger so famously put it, "Language is the house of being; in its home man [sic!] dwells."[3] But we live in particular houses; we dwell in particular languages. One speaks a language, not languages generally. Language, like number, has a peculiar ability to link the particular to the universal.

Thus, Vest is not only acting as a prophet in this volume. He is *not* simply using Wittgenstein to corral medicine and bioethics to their proper spheres of influence. The subject-object artifice that animates so much of post-Enlightenment thought is definitive neither for knowing things, nor for knowing the good. The proceduralist language of bioethics is not adequate to the richness of a life or the goods hoped-for. Language is not merely the hyphen between subject-object. It is not merely the symbol between subject and object, though it is also symbol—*sym bolon*, in Greek, bringing things together.

Vest is different from Wittgenstein, however. Wittgenstein was not concerned to plumb the depths of a polyvalent, mystical reality—perhaps he did so in his daily life, or perhaps did not live long enough to write about it. He was not himself a poet, in the literary sense. But he was living in a Europe that had lost touch with its heritage, and that resulted in the devastation of two world wars. The forms of life were gone in Europe, and Wittgenstein was left with wandering, doing work like teaching in schools for children (not universities), in hopes of finding a form of life that could reveal meaning. While not a poet, there is no doubt he lived poetically, confounding the philosophers in Cambridge.

3. Heidegger, "Letter on Humanism," 217.

Vest, however, notes that one cannot merely create for oneself a form of life. After all, we are thrown into a language not of our own choosing and thrown into forms of symbolic culture that carry long resonances in history, whether we acknowledge it or not. From whence shall come our salvation in the wake of the self-constructed, language-bearing creatures that we are? Vest's answer is not the generalized spirituality of Wittgenstein, but the long tradition of a particular religion, with particular language that draws one deeper into the mystery of being.

Vest describes the way the patristic thinkers of early Christianity formed their lives in prayer. One only has to think of Augustine's *Confessions* to see what a form of life looks like. It is a long prayer, a poetic form, that reveals Augustine's own life to himself, and allows him to see the whole of reality anew. Vest finds that forms of life are founded in the cycle of the Church year, in the prayers of the Church—pregnant in poetic language, giving form to the mystery of language and being. Language is an activity of symbolic communication, just as Divine Liturgy is the act of symbolic communication.

Divine Liturgy takes bread and wine, already having been transformed from wheat and grapes, and turns them into body and blood, bread of heaven and spiritual drink—all through the mediation of prayers, words, the Word. The Word made flesh is also already the Word breathed into creation from the beginning. The Word made flesh is also the last Word given to death to destroy death, ushering in the New Creation. The Word of creation, the Word made flesh, the Word uttered to death is the Word that turns bread into true bread—into Body—and wine into true drink—into Blood. The Word is one with God, beyond beings, and yet the Word is incarnated in the flesh of a particular man, a first-century Palestinian Jew. The Word made flesh, both universal in God and particular in a human, and made manifest in the sacrament is both symbolic signifier and that which it signifies.

But the moment one begins to theorize this, or systematize it, or conceptualize it, it begins to fail, or to produce idolatry. It is certainly true that the Christian Church has created its idols down through history. Yet, the Word is always there, to reveal, even at dusk, a form of life, rich with meaning, rich in being, and rich in the goodness of being. For words and Word are active in forming lives, in creating forms of living. It is this sort of activity—this active theology, *theo-logos*—that I had in mind when I said at the end of my book, *The Anticipatory Corpse*, "Might it not be that only [apophatic] theology can save medicine?"[4]

4. Bishop, *The Anticipatory Corpse*, 313.

Acknowledgments

THE GENESIS OF THIS book came from a doctoral thesis at the University of Nottingham under the oversight and encouragement of John Milbank. Conor Cunningham and Mikel Burley examined the thesis, offering keen feedback and criticisms. I am indebted to Dcn. Sampson Nash for reading early drafts and being a constant conversation partner throughout this process. Also, I thank Tom and Sylvia Singleton for their prayers and support, which enabled the doctorate and book to come to life. Lastly, my wife, Leah, and two boys, Jude and Lucas, patiently created space for this work. *Te amo.*

Introduction

The Puzzle of Bioethics, and Why Is Bioethics so . . . *Boring*?

BIOETHICS IS BORING?

What an odd claim for a field launched amidst the cultural revolutions of the free-wheeling mid-to-late 1960s. New medical technologies—kidney dialysis, organ transplantation, new vaccines, and more—were uncovering astonishing possibilities, forever altering the way modern, progressive societies view the body, health, and medicine. Fast forward fifty or so years, and the technological, procedural, and pharmacological capacities of medicine have further expanded at warp speed. Regenerative medicine aims to regrow damaged cells, organs, and tissues, gene therapy is taking aim at sickle-cell anemia, Ted Williams's body is cryogenically frozen, and cell phones are tracking signs of depression. What could be more exciting than navigating the ethical dilemmas surrounding twentieth- and twenty-first-century medicine?

Yet in light of this, how ironic that while receiving the 1999 Association of Bioethics and Humanities (ASBH) Lifetime Achievement Award, noted founder of bioethics Albert Jonsen asked plainly: "Why has bioethics become so boring?"[1] Jonsen has noted elsewhere that the early encounters of bioethics were daunting and full of life: "Accusing the good doctors of evil paternalism was certainly unmannerly and mean, while indicting the

1. Jonsen, "Why Has Bioethics Become so Boring?," 698n1.

scientific researchers, who were champions of American progress, for ethical misdemeanors was insulting, foolish and even dangerous."[2]

Evidently, at one point, bioethics was or seemed novel, provocative, and bold for challenging establishment science and medicine. The impact is undeniable as a new field was born, a new quasi-profession launched: the bioethicst. Science and medicine have done nothing but accelerate since the late 1960s, so why would Jonsen lament that bioethics had come to smell of boredom? Yet again, sixteen years later in 2015, Baruch Brody was awarded the ASBH Lifetime Achievement Award, and he, too, noted a stagnation in bioethics. Summarizing Brody's message, Klugman noted: "year after year at the ASBH meetings I hear colleagues say, 'There's nothing new here.'"[3]

Pensive and honest, Brody drew on Thomas Kuhn's paradigm shifts to conclude that bioethics had become normalized—the age of revolutionary bioethics had come to a close.[4] Kuhn challenges the notion that science progresses in a linear, evolutionary manner. Kuhn instead proposes that scientific knowledge grows through a series of shifting "paradigms" when the accepted "normal science" of a certain time is questioned and then overthrown by a new theory. This alternating phase of theoretical progress is an era of "revolutionary science" that then in turn becomes "normal" as the community of participating scientists test, assess, and further explore the findings of the previous revolution. Perhaps what is most striking about Kuhn's analysis is the inevitably human element in this progression as sociology, cultural and individual bias, and more play vital roles in the growth of science. Far from a crystalline, pure "other" process of logically testing and uncovering natural causes, Kuhn's paradigm draws attention to the "nearness" of human scientific knowing.

Brody's narrative hence sees bioethics emerging at the revolutionary moment when a segment of philosophy moved shop into the clinic, largely in response to the movement away from physician paternalism to patient autonomy in decision-making. As Klugman summarizes:

> *Normal bioethics* then has been the application of these insights and ideas to the world. This can be seen in the establishment of degree programs (undergraduate and graduate), professional societies, a code of ethics, even the new attestation program to professionalize clinical ethics. These are examples of a revolution

2. Jonsen, "Why Has Bioethics Become so Boring?," 691.
3. Klugman, "Bioethics," para. 5.
4. Kuhn, *Structure of Scientific Revolutions*.

> becoming normalized. Even the work in bioethics is not earth shattering.[5]

Paralleling this Kuhnian assessment, Jonsen agrees that bioethics has been normalized and accepted within the medical, scientific, and social policy worlds, but in a deeper sense he notes that "bioethics is boring because it has become too familiar, too domesticated, too much at home in accustomed places. In particular, I claim that bioethics has lived too long in the same neighborhood."[6] Jonsen then proposes two ways to shake up this overfamiliarity: bioethics should expand its "disciplinary ideas" to engage outside the "ethics neighborhood"; and secondly, bioethics "needs to travel abroad" beyond the issues and concerns particular to Americans. Through "intellectual travel" bioethics may discover new forms of thinking and "realize that contrasting customs are manifestations of distinct beliefs and values that they had not previously appreciated."[7]

On the face of it, Jonsen and Brody may be taken as veterans in their field lamenting generational changes with the nostalgia of old (academic) warriors. Jonsen even includes a lament from Alfred Lord Tennyson's "Ulysses" where the old king muses "by this still hearth" on the dullness of old age where the king "doles" out "unequal laws"[8] to those who know little of the past. Pausing here, however, how can we miss the remarkably short time period between the "early days" of bioethics in the late 1960s to the time when Jonsen and Brody describe how boring and normal bioethics has become? What does it say of bioethics that a mere fifty or so years have passed and the field has become boring? What sort of theoretical activity—while addressing deep questions on human morality and biological life—expires so quickly? What does this say about theorizing as an intellectual practice? Why did a new term, "bioethics," become popular so quickly, and why has it effectively replaced the more traditional term "medical ethics" on numerous fronts? To really get to the core questions: What do we mean by *bios*? How are we speaking of ethics? Of *ethos*? We clearly have a social and historical phenomena in bioethics. So, what do we mean by *bioethics*?

These questions lie at the heart of this book that examines in light of Wittgenstein's thinking the history, context, and interplay of methodologies in bioethics. Yet, to understand bioethics, we must look well beyond the last fifty or so years. That is to say, the recent historical events and founding figures alongside the actual topics and issues bioethics confronts present

5. Klugman, "Bioethics," para. 5 (emphasis original).
6. Jonsen, "Why Has Bioethics Become so Boring?," 691.
7. Jonsen, "Why Has Bioethics Become so Boring?," 692, 695, 696.
8. As cited in Jonsen, "Why Has Bioethics Become so Boring?," 689.

only one level of engaging bioethics. The other half of the story invites us to look deeper to a more foundational level where the metaethical questions are primary. On this foundational level, bioethics raises the questions of what it means to know scientifically and to know morally. Inevitably, these are philosophical, theological, and religious questions, and I should be clear that this latter way of approaching bioethics is more fundamental to this book than the former. In simple terms, this is a work of moral theology puzzling over bioethics. Hence, this book does not take up particular ethical questions on genetics, neurology, abortion, euthanasia, healthcare distribution, decision-making, research ethics, etc.; rather, the focus is on the *way* that bioethics as a field approaches such questions.

In particular, Kopelman's articulation of bioethics as a "second-order discipline" is helpful for identifying the approach of this book. Kopelman notes numerous approaches or understandings of bioethics: bioethics as a public discourse; bioethics as a single existing discipline—identified not by medicine or philosophy per se, but by its own content and expertise; bioethics as philosophy; bioethics as a subset of questions within medicine; and lastly, Kopelman argues for bioethics as a second-order discipline. Kopelman's proposal includes two key points: (1) that bioethics does not seek to make its subject matter "narrower" but broader in scope; and (2) that the problems of bioethics are complex and draw on different professions, disciplines, and academic fields.[9] These two points are helpful for articulating the immense spread of bioethics, and the outline and progression of this book should be taken as a proposal for one way to understand and "navigate" amidst this spread.

Moreover, this complicated landscape of bioethics struck me on a personal level when first teaching a class on the foundations of bioethics to premed students in 2013. My background in moral philosophy and the "great books" tradition would seem a fitting preparation for this arena of applied ethics, and yet what was quickly apparent is that bioethics rarely addresses the broader narrative and questions of the Western moral philosophy that experienced the crisis of the postmodern critique. The mainstream of bioethics that focuses on the Enlightenment-flavored "four principles" of bioethics hardly contemplates openly the death of God and Nietzsche's challenge to move beyond good and evil.[10] Instead, bioethics ignores such a profound moral puzzle by simply promoting a scientific and theoretical discourse that is summed up well by the most popular textbook in the field,

9. Kopelman, "Growth of Bioethics," 152.

10. See Anscombe, "Modern Moral Philosophy"; Jonston, *Contradictions of Modern Moral Philosophy*; Engelhardt, *Foundations of Bioethics.*

Beauchamp and Childress's *Principles of Biomedical Ethics, 8th ed.* (2019).[11] The more these two realizations set in, the more I began to realize the relevance of Ludwig Wittgenstein's metaphilosophy and *Weltanschauung*.[12] In particular, Wittgenstein's persistent reminders on the different modes of thinking employed in scientific theorizing and the aesthetic ethical brought about not necessarily a "new way" to understand bioethics but rather a clarified picture of the confused thinking so common in bioethics. More and more, Wittgenstein offered a deep pedagogical service in teaching premed students how to differentiate between the thin language games found in the ever-changing "code of ethics" principlist approach and the mystery (*das Mystiche*) and depth of ethics realized in the nontheoretical, religious thinking that Wittgenstein describes as a "religious point of view" (*RW*, 94). Engaging this distinction with my students clearly does not lead to a simplistic new "theory" in bioethics as much as it changes the "form" of thinking employed in bioethics and the expectations we have for positive, scientific knowledge in the sphere of ethics. Even more, this distinction helps to explain the boring normalizing that has beset bioethics within a mere fifty years; the "revolutionary" language has been spoken; conferences have convened, applying the principles to all the popular topics; the articles and books have now been written. In short, the theoretical language game of bioethics has been played.

Part I: Dilemmas in the Spirit of Scientific Theorizing

This book tells that story of bioethics as language game in three main parts. Part I introduces the recent birth and phenomenon of bioethics as an applied secular philosophy. Chapter 1, "A Secular Phenomenon: The Birth of Bioethics after God," probes beyond the standard narrative of bioethics arising out of twentieth-century human subjects research atrocities—most notably the Nazi Holocaust—or of bioethics meeting new scientific-medical technological advances (and shortages), or of bioethics correcting imbalanced physician paternalism. Undoubtedly these narratives are critically important, and yet they tell only part of the story. Below the spheres of the cultural and social, we can and should discern important theological and ontological differences at play within the birth of bioethics. At the most fundamental level, bioethics emerges as a dilemma or puzzle: When

11. See Eckenwiler and Cohn, *Ethics of Bioethics*; Jones, "Unholy Mess"; Lee, "Problem of 'Thick in Status, Thin in Content'"; Arras, "Hedgehog and the Borg."

12. Monk, *Ludwig Wittgenstein*, 516.

the modern and postmodern worlds have no firm foundation for ethics, how do we proceed in an *applied* field of ethics such as bioethics? Who is most authorized to claim bioethical expertise? Who possesses the proper jurisdictional authority, and to whom should or ought we affix the title of "bioethicist?" Many of the earliest figures in bioethics were theologians who shortly turned—implicitly and explicitly—from the language and methods of traditional theology towards a more socially relevant posture as bioethicists. This radical move affirmed the inauguration of the pragmatic, professionalized, principlist language of bioethics.

In chapter 2, "Introducing Wittgenstein: Prophet of Modern Theorizing," we turn to Wittgenstein's life and thinking, which poignantly engage these sorts of questions. In a real way, Wittgenstein prophetically anticipates the struggles to ground bioethics amidst morphing language, and he has much to offer us amidst his own early puzzles over language and philosophy grounded in scientific or theoretical thinking. Chapter 2 lays key groundwork by arguing for the right way to understand the relationship between the "early" and "later" Wittgensteins; while not denying the different emphasis between the early Wittgenstein on logic and reality-as-representation and the later Wittgenstein on the limits of language, what bridges both Wittgensteins is a metaphilosophy, or unity of purpose, in seeking to uncover the errors of modern philosophy so "preoccupied with the method of science" that lead philosophers "to ask and answer questions in the way science does" (*BB*, 18).

Part II: Critiquing Scientific Theorizing and Scientific Ethics

Chapter 3, "Collapsing Theory: The Impulses of Scientific Thinking," begins the turn from *identifying* these two parallel modern dilemmas—early Wittgenstein and bioethics—to *examining* and *critiquing* these impulses in scientific theorizing. Wittgenstein's critique hinges on a use of language in a mode of scientific theorizing that leads to a confused and faulty metaphysics. In particular, Wittgenstein rejects metaphysics such as Kant's transcendental idealism and instead employs language as a practice and therapy for revealing our immanent, finite knowing as fundamentally contingent. Grammar or the rules of language for Wittgenstein do not provide a pseudoontology but rather a way of showing the limits of our language to define or explain *das Mystiche*. Wittgenstein hence offers a way to change the mode of ethics from a scientific and self-referential game grounded in immanent human knowing to a more authentic ethics without (theoretical) philosophy.

Chapter 4, "Bioethics as a Self-Referential Language Game," exemplifies this Wittgensteinian thinking in the sphere of bioethics where modern science routinely expands beyond empirical data claims into philosophical and theological territory. Nuclear physicist Alvin Weinberg's notion of "trans-science" illustrates this debate over transgressing the limits of scientific knowing, and when the bounds of scientific knowing are this fluid, a system of language emerges inevitably to provide an ordering structure. Julian Savulescu in particular illustrates a self-referential, utilitarian language game in bioethics, and while Savulescu ironically presents a scathing critique of principlist bioethics—revealing the lack of grounding for "moralist" terms such as autonomy and consent—he fails to note well his own assumptions regarding how to rank the scale of goods in his own utilitarian methodology. In the end, after unpacking Savulescu's framework of enhancement and moral criteria for defining the good, what most emerges is nothing less than an implicit fidelity to the Nietzschean posthuman god.

Part III: Escaping the Ethical Fly Bottle

The final part of the book takes a redemptive turn to examine the strong family resemblance between Orthodox bioethicist H. T. Engelhardt and Wittgenstein in chapter 5: "Pedagogy in the Ruins: Why Procedural Bioethics Matters." Engelhardt is well known within bioethics primarily as a philosopher of difference, one who points out the fundamental disconnect between the claims of optimistic, content-full bioethics and the realistic ethical confusion that emerges from today's pluralistic, postmodern landscape. Similar to Wittgenstein's practices with language that reveal limits of immanent knowing, Engelhardt traces through secular bioethical reasoning the thin possibilities of consensus by permission. Far from exulting in the thin agreements that arise from secular ethical procedures, Engelhardt "laments" the limits of secular bioethics amongst "moral strangers"[13] in hopes of directing others to the content-rich possibilities of religious—and more specifically, traditional Christian—bioethics. Hence, while both philosophers employ a stringent critique and philosophy of difference, both do so as a pedagogical practice, calling on us to discern the lamentable limits of philosophy and bioethics in the ruins of postmodernity. While Wittgenstein advances tangible forms of life (*Lebensformen*) in place of abstract, scientific theorizing, and while Engelhardt advocates for a libertarian, procedural bioethics, both philosophers seek these aims as therapeutic ways of life to

13. Engelhardt, *Foundations of Bioethics*, xi, 6–7.

direct us away from idols of the secular postmodern mind and towards the richness of transcendent mystery.

Chapter 6, "How Sacred Cosmology Saves Ethics," draws together the themes of the book, moving past a Wittgensteinian critique of theorizing to ask what this approach offers more constructively to ethics and bioethics. Admittedly, consistent with Wittgenstein's metaphilosophy, this conclusion offers neither a new theory nor a systematically constructive bioethical paradigm. Rather, Wittgenstein's constructive turn is particularly relevant to the foundations of ethics and bioethics as a first-level metaethics. This first-level, constructive turn brings to light Wittgenstein's "religious way of thinking" that embraces an anthropology and moral epistemology very much at odds with the modern assumptions of Cartesian personhood and theoretical ethical knowledge. In sum, this chapter seeks to change the *form* of ethics and bioethics by exploring three themes in a Wittgensteinian key: the first is Wittgenstein's religious point of view (*RW*, 94) that recasts ethics as a mode of thinking and being in the pursuit of sacred and transcendent reality; the second explores ethical thinking as a form of apophatic and embodied knowledge; the final theme seeks to show the contrast between scientific thinking in parts (theoretical ethics) and cosmological thinking (theological ethics). Without a doubt, this final chapter steps far away from the *ordo* of secular bioethics that functions notably through moral *techne*; rather, the final aim here is to present an ancient Christian mode of ethical thinking that maintains the unity of the heavens and earth, of spirit and matter, of *vios* and *bios*.

Abbreviations

BB *The Blue and Brown Books.* Oxford: Blackwell, 1958.

CV *Culture and Value.* Edited by Georg Henrik von Wright & Heikki Nyman. Translated by Peter Winch, Oxford: Blackwell, 1980.

LE *Lecture on Ethics.* 1st ed. Established Text of the Lecture, *MS 139b Normalized*, edited by E. Zamuner et al., 42-51. Oxford: Wiley & Sons, 2014. Pages cited here follow the original page turns cited in Zamuner et al. in square brackets.

MS refers to all unpublished material compiled in the Wittgenstein *Nachlass.*

OC *On Certainty.* Edited by G. E. M. Anscombe & Georg Henrik von Wright. Translated by Denis Paul & G. E. Anscombe. New York: J. & J. Evanston, 1969.

PI *Philosophical Investigations.* 4th ed. Edited by Peter M. S. Hacker and Joachim Schulte, translated by G. E. M. Anscombe et al. Oxford: Blackwell, 2009.

PP *Philosophy of Psychology—A Fragment* [previously known as "Pt. II"]. In *Philosophical Investigations*, 4th ed. Edited by Paul M. S. Hacker and Joachim Schulte. Translated by G. E. M. Anscombe et al. Oxford: Blackwell, 2009.

PR *Philosophical Remarks.* Translated by Raymond Hargreaves & Roger White. Oxford: Blackwell, 1975.

RF "Remarks on Frazer's Golden Bough." In *Philosophical Occasions*, edited by James C. Klagge & Alfred Nordmann, translated by Rush Rhees, 115–55. Indianapolis: Hackett, 1993.

RW *Ludwig Wittgenstein: Personal Recollections*. Edited by Rush Rhees. Oxford: Blackwell, 1981.

T *Tractatus Logico-Philosophicus*. Translated by Charles K. Ogden. Mineola, NY: Dover, 1999.

VC *Ludwig Wittgenstein and the Vienna Circle: Conversations Recorded by Friedrich Waismann*. Edited by Brian McGuinness. Translated by Joachim Schulte & Brian McGuinness. New York: Barnes and Noble, 1979.

Part I

Dilemmas in the Spirit of Scientific Theorizing

1

Introducing Bioethics: Applied Secular Philosophy

1. History of a Neologism

Nietzsche's admonition that all of philosophy is in the end "involuntary and unconscious memoir" should be considered in relation to all bioethics texts. The choice of an impersonal narrator is simply an attempt to put on a mask that conceals how personal the narrator truly is. —Tod Chambers[1]

As with many developments in language, the birth of the term *bioethics* is instructive. According to Warren Reich, "the field of bioethics started with the word bioethics because the word is so suggestive and so powerful; it suggests a new focus, a new bringing together of disciplines in a new way with a new forum that tended to neutralize the ideologic slant that people associated with the word ethics."[2] Further, this "heuristic"[3] neologism

1. Chambers, "Virtue of Attacking the Bioethicist," 286.

2. Reich, "How Bioethics Got Its Name," 7.

3. "A new word often allows us to name elements of reality in a way that conveys new control over our cultural environment. It is often not the precision of a word that is the source of its power and usefulness. In fact, it is often the imprecision, the lack of clarity, that allows us to name and bring together at one time many areas of interest. An apt word can assemble a rich set of images and meanings and thus help us to see relations between elements of reality that were previously separated in our vision and

carries a concrete history in time—the end of the 1960s—and place, namely three nascent academic centers: The Institute of Society, Ethics and the Life Sciences (The Hastings Center, 1970), the Kennedy Institute of Ethics (1971), and The Society for Health and Human Values (1969).

The founders of these centers were varied but all carried notable theological roots. Alongside Dan Callahan (former editor/contributor to the Catholic journal *Commonweal*) and Robert Veatch (PhD in Religion and Society), protestant theologians Paul Ramsey and James Gustafson were influential in the launching of The Hastings Center. Formed within Georgetown University, The Kennedy Institute of Ethics benefitted notably from André Hellegers's vision and the support of Eunice Kennedy Shriver and the Joseph P. Kennedy Foundation. By medical training, Hellegers was an obstetrician and gynecologist, but given his interests in philosophy and theology, Hellegers found "the atmosphere of the Jesuit university congenial."[4]

In 1971, the Kennedy Institute's first two research scholars were Mennonite theologian LeRoy Walters and Warren Reich, formerly a moral theologian at Catholic University. Other theologians soon followed, including Charles Curran, Richard McCormick, Gene Outka, John Connery, and in 1975 James Childress, also a theologian by training, came from the University of Virginia.[5] In 1998, The Society for Health and Human Values (SHHV) merged with the Society for Bioethics consultation to form the American Society for Bioethics and Humanities (ASBH). As with the Kennedy Institute and the Hastings Center, ASBH's roots were theological and religious as the SHHV first originated in the mid-sixties as a collaboration of Methodist and Presbyterian churches' efforts in medical education and theology.

The role of theology and theologians will surface again, but some further history is helpful at this point. Within Jonsen's popular narrative of bioethics (*Birth of Bioethics*, 1998), the term bioethics had a "bi-located birth" between 1970 and 71, with two notably different visions for the future

thought of only as disparate. . . . This has been the case with 'bioethics.' . . . The word 'bioethics' [has done] brilliant service in bringing together a wide cluster of important cultural concerns. The term [is] profoundly heuristic" (Engelhardt, "Foreword," vii).

4. Jonsen, *Birth of Bioethics*, 22. One of the significant projects to emerge from the Kennedy Institute for Ethics came from Hellegers's push for Reich to organize and map this newly emerging field. Reich accepted this challenge, and with the financial support of the National Endowment for the Humanities published a four-volume *Encyclopedia of Bioethics* in 1978. For an emerging field so recently named "bioethics," it is remarkable to see the knowledge and structure of bioethics organized so quickly as an encyclopedia.

5. Jonsen, *Birth of Bioethics*, 23.

of bioethics.[6] Jonsen points to research oncologist Van Rensselaer Potter as the first who began writing of bioethics in a 1970 article[7] that was followed by a book in 1971, *Bioethics: Bridge to the Future*. Potter's vision should be seen as an extension of Aldo Leopold's land ethic to include all elements of human environment, and the opening of *Bioethics: Bridge to the Future* quotes from Leopold on three possible modes of ethics: the first between individuals (e.g., the Decalogue); the second between individuals and society (e.g., the Golden Rule); and the third which consists of an extension of this sequence to ecological considerations. Bioethics, as Potter envisioned it, should focus on much more than medical ethical questions by rather taking into consideration all biological, communal, relational issues of man within the ecology of the natural world. Pursuing such a harmony, Potter sought to unite the "two cultures" of science and the humanities. This involved an expansion beyond both mechanistic interpretations of biology and immaterial views of ethics. Biology and wisdom hence are intimately connected for Potter, and he did not shy away from connecting these pursuits to the survival of humanity currently hell-bent on technologically and economically consuming the finite resources of the physical world.[8] Potter draws together observations from thirty years of oncology research alongside environmental philosophy and spiritual-scientist philosophers such as Teilhard de Chardin. For many familiar only with bioethics literature from the last decade, it is likely that Potter's "Bioethics Creed for Individuals" will seem entirely disjointed from contemporary bioethics. Potter's credo integrated five statements of belief and commitments regarding: (1) personal acceptance of global, ecological crises; (2) mankind's role in these crises; (3) the uniqueness of the individual-society relationship; (4) the inevitability of human suffering with the commitment to not passively accept further suffering caused by humanity; and (5) acceptance of finitude and finality of life as necessary to further generational life. In short, Potter is an early prophetic voice in bioethics, calling for a clear integration of philosophical and theological thinking, a comprehensive moral methodology far from reductivist ethical theorizing:

> [Bioethics] would attempt to generate wisdom, the knowledge of how to use knowledge for social good from a realistic knowledge of man's biological nature and of the biological world. To me, a realistic knowledge of man is a knowledge that includes his role as an adaptive control system with built-in error

6. Jonsen, *Birth of Bioethics*, 27.
7. Potter, "Bioethics, The Science of Survival."
8. Potter, *Bioethics*, 1–29.

> tendencies. This mechanistic view, which combines reductionist and holistic elements, would be totally incapable of generating wisdom unless supplemented with both a humanistic and an ecological outlook . . . the present world is dominated by military policy and by an overemphasis on production of material goods. Neither of these enterprises have given any thought to the basic facts of biology. An urgent task for Bioethics is to seek biological agreements at the international level.[9]

Within Jonsen's narrative, André Hellegers was the second figure laying claim on the term bioethics, and his rival use of the term has clearly won the day. Rather than viewing ethics as a broad, inclusive term that addresses ecology, biology, philosophy, and spirituality as did Potter, Hellegers and others at Georgetown saw ethics as "a rigorous examination on the grounds for moral norms."[10] The Hellegers/Georgetown model became such the assumed framework, moreover, that many of the early essential works in bioethics—including Reich's *Encyclopedia of Bioethics* (1978), Daniel Callahan's "Bioethics as a Discipline" (1973), LeRoy Walters's *Bibliography of Bioethics* (1975), and Beauchamp and Walters's *Contemporary Issues in Bioethics* (1978)—all but fail to mention Potter's vision for bioethics.[11] The difference between the two visions is strong, and it is easy to focus on Potter's vision as a cosmological or global ethics while Hellegers/Georgetown focuses on narrow medical ethics, but as Reich demonstrates, the matter is more complicated. First, Potter used the term "global" bioethics in an ambiguous way that could:

> (1) relate to or involve the whole earth: a worldwide ethic for the good of the world; (2) entail the comprehensive inclusion of all ethical issues in the life sciences and health care (both the "biomedical" and the "environmental" issues of this classic debate); and (3) utilize a comprehensive vision of methods for approaching these issues: expansively incorporating all relevant values, concepts, modes of reasoning and disciplines.[12]

Potter seems to have emphasized all three of these aspects of global bioethics *as* bioethics at different times, while the Hellegers/Georgetown model unmistakably treated medical ethics *as* bioethics, narrowing the focus considerably. Further, Potter's experience in oncology no doubt influenced his desire to look for preventative measures, environmental health,

9. Potter, *Bioethics*, 26.
10. Jonsen, *Birth of Bioethics*, 27.
11. Reich, "Word 'Bioethics,'" 24.
12. Reich, "Word 'Bioethics,'" 24.

agricultural policies, as well as global health education, and, broadly speaking, these aims align with Hellegers's work on "the worldwide disequilibrium between the powerful and the powerless" as well as global human infertility amidst social and economic problems in developing countries.[13] Reich argues that Hellegers's overarching message addressed much more than surface, procedural questions and that the developing ways of formulating bioethics debates "begged the question of why we are in medicine in the first place." In short, "implicitly and indirectly," Hellegers acknowledged "shortcomings" in the Georgetown model of bioethics even while "enthusiastically taking part in it."[14] No doubt in part due to the strength of the Georgetown University name and the funding grants Hellegers successfully secured, bioethics became almost entirely associated with dilemmas in bio*medicine*.

Earlier than Hellegers and Potter, however, we may see a use of the term bioethics in a 1927 publication from Protestant pastor and philosopher Fritz Jahr, "Bio-Ethik. Eine Umschau über die ethischen Beziehungen des Menschen zu Tier und Pflanze" ("Bio-Ethics: A Review of the Ethical Relationships of Humans to Animals and Plants"). Jahr's message could have been taken as yet a third possible direction for bioethics. Hellegers, Potter, and Jahr all sought a renewed vision "for ethics and morality in times of new technologies and changing cultures,"[15] according to Sass, but Jahr differs from Potter and Hellegers in following a distinctly theological and philosophical path to address the ethical implications inherent within all forms of life. Sass transposes Jahr's language to articulate a view of "*Bioethik*" as "the systematic study of human conduct in the area of the life sciences and the personal, professional, and public moral commitment and conduct towards all forms of life, in as far as this conduct is examined in the light of moral values and principles."[16]

This transposed language is familiar to contemporary bioethics, and yet what emerges primarily in Jahr is a close relationship to nineteenth-century thinking in the life sciences that could not be more distant from contemporary bioethics. Sass points to figures such as Wilhelm Wundt (*Lectures on Human and Animal Psychology*, 1863) and Theodor Fechner (*Nanna: Über das Seelenleben der Pflanzen* [*Nana: Or the Soul Life of Plants*], 1848; *Elemente der Psychophysik* [*Elements of Psychophysics*], 1860) as examples of a turn away from Cartesian body-soul dualism,

13. Reich, "Word 'Bioethics,'" 25.

14. Reich, "Word 'Bioethics,'" 28.

15. Sass, "Fritz Jahr's 1927 Concept of Bioethics," 280.

16. Sass, "Fritz Jahr's 1927 Concept of Bioethics," 282.

focusing instead on teleological aims and interactions "of and between living and sensing environments and living and sensing beings."[17] Alongside biophysics and biochemistry, Rudolf Eisler (*Leib und seele: Darstellung und Kritik der Neueren Theorien des Verhältnisses Zwischen Psychischem und Physischem Dasein* [*Workings of the Soul: Ideas Towards an Organic Psychology*], 1909) proposed "Biopsychics" as a way to comprehend the willing activity of all forms of life.

Similar to these Austro-German figures, one may think of the French "spiritualist" Félix Ravaisson (1813–1900), who traced an ontological phenomenon of "habit" as a way to grasp the intimate relationship between being and virtue within material nature. Namely, as things or subjects change, the familiarity of habits diminishes resistance, while passivity (non-habitual activity) increases resistance. This seemingly simple observation in nature reveals much as Ravaisson seeks to further Aristotle's search for "the immaterial reality of individual, particular things."[18] The goal is to see in all organic nature the "essence" of these things in their respective "*noien* thought," and in this identification of spiritual or intellectual intuition within all organic-sense activity, the goal is to uncover the genuine unity of all being.[19] Within such a framework, any hard distinction between ontology and ethics is meaningless as the essence of a thing, in one way, inherently manifests or reveals its virtue, and, conversely, the virtue of a thing directly influences the ontological. Jahr's *bio-ethik* hence may be seen as a fusion of biology and psychology, leading Jahr to gesture towards examples as varied as St. Francis of Assisi's love for animals and Rousseau's enthusiasm for nature. Such a context for bioethics is certainly far from contemporary bioethics literature, but Jahr emerged from this context into more familiar terrain by promoting Kant's categorical imperative into a "bioethical imperative" founded on the a priori premise that all of creation is sacred. This bioethical imperative draws upon a love and compassion for all things, the Golden Rule (Matt 7:12; Luke 6:31), and inherent moral duties to oneself under the banner of "personal health responsibility."[20] Jahr draws heavily from Kant and yet differs from much of the Kantianism in contemporary bioethics

17. Sass, "Fritz Jahr's 1927 Concept of Bioethics," 280.

18. Carlisle and Sinclair, "Introduction and Commentary," 2.

19. Clare Carlisle and Mark Sinclair quote Henri Bergson on this point: "The unity joining beings to one another, the unity of a thought that we see, from inorganic matter to the plant, from plant to animal, from animal to man, until from concentration to concentration we should end in divine thought, which thinks all things in thinking itself" ("Introduction and Commentary," 2).

20. Sass, "Fritz Jahr's 1927 Concept of Bioethics," 283–89.

by embedding his Kantian-spiritualist methodology within an ontological unity and continuum of all things.

When comparing and contrasting Hellegers, Potter, and Jahr, a similarity arises between Jahr and Potter's broader, integrative bioethics, respectively, and yet both visions have fallen to the wayside. An ethics drawing together cosmic harmonies of the quadrivium has hardly emerged from the nascent field of bioethics. Hellegers's biomedical focus for bioethics, rather, has won supremacy as evidenced by Potter's later retreat to rebrand under "global bioethics."[21] Doubtless there are additional layers to understand within the legacy of the bioethics neologism, but it is critical here at the start to note the secular and theoretical order that accompanied Hellegers's use of bioethics, regardless of his intentions. Key to the argument of this book is the secular dilemma within bioethics that runs parallel to the dilemma of modern moral philosophy itself, and this dilemma is well documented through figures such as Anscombe from two decades past—the dilemma of moving forward in modern moral philosophy as though analysis of moral concepts and presuppositions can take place when the fundamental existence of those concepts, presuppositions, and ways of life remains in question. Such a maneuver reduces modern moral philosophy to an examination or analysis of justice or virtue in light of certain "facts" or realities that remain unexamined, and the whole moral endeavor then operates on the level of thin theoretical analysis while suspending judgment on the "facts" or realities undergirding such analysis. In other words, modern moral philosophy and bioethics are postured as tools of analysis amidst a neutral environment with suspended values. Hence Anscombe notes in such arguments that "'wrong' of course is explained as meaning 'morally wrong' and all the atmosphere of the term is retained while its substance is guaranteed quite null."[22] To this we may consider Engelhardt's description of what occurred when bioethical visions such as Jahr's and Potter's were abandoned:

> For the most part, the term bioethics has been taken to identify the disciplined analysis of the moral and conceptual assumptions of medicine, the biomedical sciences, and the allied health professions. As such, it has become a special area of philosophy of ethics, even though all of its practitioners have not been formally trained in either area.[23]

21. Potter, *Global Bioethics*.
22. Anscombe, "Modern Moral Philosophy," 17.
23. Engelhardt, "Foreword," x.

Enframed within secular and theoretical modern moral philosophy, bioethics proceeds not due to its "factuality" within a coherent way of life but due to its "acceptance," an acceptance based on what Tom Koch describes as bioethics's "foundational myth."[24] The myth more or less follows Jonsen's narrative, as described above, that from the 1960s and 1970s bioethics arose as a capable replacement for a traditional ethics of medicine that failed to keep up with new technological and scientific realities. Bioethics then exists primarily as a set of analytical tools "accepted" or selected by the powers that be, and the context for this acceptance is a "neoliberal, modernist order." By modernist, Koch means turning from "worship" of past traditions and authorities to a future pursuit of projected goods, pleasures, and freedoms. By neoliberal he means an order "in which consumerism and transactional thinking are advanced, in the name of individual autonomy, over values of community and communal responsibility. The result is not morally rich but instead ethically blank, a bookkeeper's recording rather than a moralist's recording."[25] Bioethics in this way becomes a mere transaction and interchange of theories intended to guide and regulate scientists, physicians, and others engaging matters of *bios*.

Further supporting the foundational myth of bioethics are particular roots within Western philosophy that assert simplistically that reason guides the will towards a set of applicable moral laws that are "out there." Mark Johnson calls this "the Moral Law folk theory,"[26] and Johnson's poster child for this is Immanuel Kant and his arguments for the autonomous, rational individual who may through reason perceive the moral law.[27] What emerges is not necessarily Kant himself but rather a Kantian move to place moral authority in autonomous reason removed from cultural, social, and religious traditions. Such a narrow approach to bioethics as a neutral, theoretical analysis ignores numerous layers of complexity, not to mention the fundamental insight from figures such as Hegel, Wittgenstein, and Charles Taylor that there is no direct line of rationality between the transcendent and the immanent or mundane. The complexity of cultures, psychology, religious belief, etc. creates a scenario that calls for much more in bioethics than simplistic Moral Law folk theory or thin appeals to the four principles of autonomy, beneficence, nonmaleficence, and justice. Appealing to these

24. Koch, *Thieves of Virtue*, 9.

25. Koch, *Thieves of Virtue*, 9.

26. Johnson, *Moral Imagination*, 4.

27. Regarding Johnson's Moral Law folk theory, Koch states: "Moral reasoning is thus principally a matter of getting the correct description of a situation, determining which moral laws pertain to it, and figuring out what action that moral law requires for the given situation" (*Thieves of Virtue*, 10; see also 10–15).

few theoretical tools assumes some "pure humanity" existing within a supposedly neutral moral space or domain, and, hence in effect, bioethics seeks to exist within a secular sociology when what is forgotten is that "the secular is a domain that had to be instituted or *imagined*, both in theory and in practice."[28] Every person physically stands *somewhere* as every person necessarily is begotten by parents (even if the parenthood is purely biological). To suppose otherwise is farcical, and so, too, secular, autonomous rationality is not the summit of an ascending moral chain of rational being, finally free from cultural, social, and religious trappings.

In short, from the start it is critical to see bioethics arising within the context of a particular phenomenon—a historical, cultural, and philosophical phenomenon. Historically, bioethics emerges from an era of remarkable upheaval in the sixties and early seventies in terms of new medical technologies and possibilities. This era experienced seismic shifts and upheavals ranging from the Vietnam War to Vatican II to the Civil Rights movement and more. Academic centers and institutions of bioethics sprang up amidst this era of change, launching a new field under the crafty neologism "bioethics." Medical ethics remains a term employed by some who wish to distinguish from the ambiguity of "bioethics," and yet the very ambiguity of bioethics has allowed for a gathering of institutional and academic resources around this new specialty imbibed with new energy. Culturally, medicine and healthcare remain at the heart of contemporary political deliberations, public policy, technoscientific research, and more. Hellegers's vision of bioethics focused on this thriving technological biomedical era has succeeded over Jahr and Potter's cosmological and spiritual visions of bioethics. Thirdly, and most importantly for our purposes, bioethics should be seen as a late post-Enlightenment phenomenon of secular philosophy. In this way, the story of bioethics is a retelling of the transition in modern moral philosophy towards moral theorizing and principlism as opposed to foundational moral *theoria*. As the ancient schools of philosophy attest, or as a traditional religious bioethics exemplifies, epistemological ways of knowing are united to our way of being, and such a being is expressed ethically. The transaction and interchange of theories in bioethics reflects an ideological marketplace that presupposes philosophy as a secular, neutral, cosmopolitan endeavor. Hence bioethics is confronted with the predicament of the modern secular order where philosophy seeks (1) to exist in a mode of pseudoneutrality, enacting a "view from nowhere," and (2) to further this philosophical posture by distancing theory from praxis.

28. Milbank, *Theology and Social Theory*, 9 (emphasis original).

Inevitably, this thin, transactional approach to moral theorizing within the modern secular order takes bioethics ever more into the sphere of biopolitics where the greatest import of the ethical is the political, a theme that will be explored shortly. For now, however, a fitting parallel to summarize the phenomenon of bioethics can be drawn from Hervé Juvin's depiction of a great novelty in the twenty-first century: the creation of a new body. Juvin traces the effects of raw capitalism within medicine, and this nexus and malleable marketplace enacts a redefinition of what it means to be human. The body is no longer the embodied soul and place of physical efforts and toil with the soil, burdens, and numerous struggles to live. The new body is not something that decays amidst the natural journey from birth to death; it is rather a novel entity that seeks fundamentally to resist "need, suffering, and the effects of time." The body is a thing that resists nature and external destiny as it seeks to will a new destiny:

> Life is managed. Life is produced, and is constructed in the world of technology and method to such an effect that nature is a taskmaster no longer, having become a mere argument; to such effect too that after a world of nature and destiny, we are discovering the world of choice: the world of the market.[29]

According to Juvin, the "absolute novelty" of the early twenty-first century is that we face the challenge of "managing" a body as a "product" that we have created. Amidst this new vision of the body—and a medical practice that ever challenges the line between therapy and enhancement—transactional, theoretical bioethics is well suited to adapt to and to promote whatever desired end goals arise within these new parameters. Ethical language and ethical theories are malleable to the desires or "rules" of the game at hand, as Wittgenstein would note. The end goal may be a transhumanist vision of perpetual life, a view of eugenics for all that effectively removes all disabilities, or the end goal may be Juvin's prophecy of the "cult of the body beautiful" that creates a new morality of health applied in fitness centers, dieting, analysts, cosmetic surgeons, etc.[30] The possibilities are manifold, and the language of bioethics becomes all the more critical as the principlist model of autonomy, beneficence, nonmaleficence, and justice morphs to fit various world pictures as defined by the one who speaks. When such principles and language float freely, dystopian and violent goals for human life need not worry about bioethical safeguards and concerns; bioethics will be well suited for theoretical adaptation as a defense of those very goals.

29. Juvin, *Coming of the Body*, x.

30. Juvin, *Coming of the Body*, 95.

2. The Biopolitical Question: Whose Language? Philosophers, Priests, or Bioethicists?

> *The supervision of [propagation, births and mortality, the level of health, life expectancy and longevity] was effected through an entire series of interventions and regulatory controls: a bio-politics of the population. The disciplines of the body and the regulations of the population constituted the two poles around which the organization of life was deployed.* —Michel Foucault[31]

All books on ethics or bioethics inevitably begin with a number of core philosophical assumptions, and a discerning reader will quickly ask for clarity regarding the ground rules. When first opening a popular textbook such as Lewis Vaughn's *Bioethics, Principles, Issues, and Cases 4th ed.*, the reader will learn of three basic aspects of bioethics: (1) that bioethics is identified first and foremost as an offspring discipline of philosophical ethics; (2) that grave moral issues of life, death, health, illness, pain, relief, hope, and despair are at stake; and (3) that bioethics holds out the possibility of answers for such questions.[32] This seems well and fine, but how is the discerning student to understand these three tenets in light of philosophy's modern history, methods, and metaconvictions? Hence the fundamental historical question identified above: How does bioethics align within the recent history of modern moral philosophy?

With this question before us, the next step is to examine the context in which this question swirls. Navigating the context of this question inevitably brings to light the possible methodologies for answering the question, and with remarkably popular (and profitable) textbooks like Vaughn's, both context and methodology are remarkably neglected. Amidst the vast history of moral philosophy in the West and East, why must the primary choices fall between derivatives of Kantian duty, utilitarianism, ethics of care, and more? Why is theory so prominent in bioethics over other systematic, analytic, or existential philosophical methods? As another way of asking this: What metaphilosophical assumptions undergird bioethics? Bioethics is committed to the pursuit and wellbeing of life (*bios*), but is it too much to ask "Why life?" or "Whose life?" Nietzsche, for one, is hardly a philosopher widely quoted in bioethics, and yet at the very least bioethics should recognize the need to confront Nietzsche's particular views of sickness, suffering, desolation, indignity, self-contempt, and more.

31. Foucault, "Right of Death and Power Over Life," 262.

32. Vaughn, *Bioethics, Principles, Issues, and Cases*, 3.

To be sure, these critical questions belong within the language of philosophy, which is the starting point for most bioethics textbooks, and yet a more expansive consideration for the language and issues of bioethics is needed. As we will see, bioethicists do not speak solely in the language of philosophers, pondering and reflecting upon the good by seeking to know what is right and just.[33] In addition to philosophy, bioethics engages religion as well as law and policy in particular, each sphere with its particular language system. In the language of religion, priests will offer degrees of imperative instruction through canonical teaching or through the manualist tradition present within Roman Catholicism until Vatican II.[34]

The language of secular, theoretical bioethics, however, is far removed from religious bioethics, and indeed the "Moral Law folk theory" described earlier would have us think that autonomous reason guides the will to the moral law apart from any religious thinking or practice. Within the Moral Law folk theory, secular and rationalist assumptions are the *de facto* grounding, and any religious bioethics is necessarily irrational, even superstitious, and hence an aberrant presence amongst theories that are acceptable at the table of cosmopolitan deliberation.

In other words, Vaughn's textbook and the Moral Law folk theory stand up a standard narrative that bioethics is a neat subset of applied philosophy. To the contrary, what this narrative most misses is the necessary

33. Some may reasonably contest that the methodology of bioethics has much at all in common with ancient philosophy. Plato, Aristotle, and many others naturally speak the language of philosophers—inquiring after the unity, difference, and meaning in various forms of *bios*—and while their ascetic and pedagogical approaches are far from today's bioethics, this is a finer distinction on "what is philosophy" than exists for bioethics today. The point for now is to note that bioethics sees itself as existing broadly "within philosophy" as a subgroup of applied ethics.

34. McDermott, "Collapse of the Manualist Tradition." The language of priests may follow a distinct authority structure such that, for example, within the ancient Christian tradition, there are church canons that provide an authoritative touchstone. This is not an authoritarian structure that ignores pastoral application within the spiritual life of Christians under a spiritual father, nor does it reduce the language of priests to a collection of holy principles, rules, or conclusions. In the ancient Christian tradition, ethics cannot be considered apart from seeking "the kingdom of God and His righteousness" not of "this world" (Matt 6:33; John 18:36).

Further, for Muslims, Islamic bioethics takes *shari'a* and the jurists' *fatwas* as authoritative, while Jewish bioethics draws upon the *halakhah* as rabbinic law. It would be advantageous in some ways to understand bioethics if the language and system of philosophy or religion were the consistent norm. Indeed, when one proceeds "to do" bioethics within the structure of a Jewish, Christian, or Muslim ethics, one can expect many shared presuppositions and practices relating to ecology, health, and life. Mainstream contemporary bioethics, however, speaks through an "impersonal narrator" and often claims jurisdiction beyond a religious or theological bioethics, seeking to impart a "pure" philosophical reflection on moral issues.

examination of bioethics in light of a produced and institutionalized formal rationality—namely, bioethics as biopolitics. To understand the critical role of biopolitics, we need to look only to one of the most prestigious voices in bioethics, the United States National Bioethics Advisory Commission (NBAC). The NBAC assembles for moral deliberation, advising the president and nation on matters of utmost importance, and yet the sentiments of this commission inevitably change every four to eight years as a new president enters the White House. What does this say of morality and ethics when such a leading voice in bioethics likely changes its positions as political regimes change? Idealistically speaking, the commission may be viewed as a platform for philosophical debate where robust ethical differences are aired, and yet the commission in reality remains a body of appointed scholars, lawyers, and physicians within a particular political regime. The stated goal of each commission is to speak democratically, and yet what does it mean to pursue ethics democratically? The NBAC may speak in the *lingua franca* of philosophical principles, but the real system and workings of power within the NBAC—and, by extension, much of bioethics on the whole—incorporate social, cultural, institutional, and political power dynamics. Modern bioethics, in other words, assumes biopolitics.

What is biopolitics? As a way of answering this, John Evans points us towards the system of language in bioethics that is best understood by following the "cultural production" and "institutionalization" of key bioethics texts and documents.[35] An example of this can be seen in the dialogue on human genetic engineering (HGE), a dialogue that has changed significantly since the 1960s when—as Robert L. Sinsheimer has pointed out—the debate was quite open-ended regarding what constitutes the perfection of species. By the 1990s, however, the "ends" that directed the debate were narrowed and directed to operate within "a very constrained list of universal, commensurable ends that [had become] institutionalized by the dominant profession in the debate."[36] Not surprisingly, autonomy arose as the primary end goal directing the HGE debate, leaving those yet to be born ironically as the presumptive agents needing to give informed consent for any genetic modification. Sinsheimer himself advocated a "new eugenics" that evokes immense controversy, yet such controversy stems from robust or "thick" end goals for what constitutes human flourishing. For now, what matters most is not Sinsheimer's particular argument for

35. John Evans uses human genetic engineering in particular as a lens to trace the sociological power structures in bioethics. In the section below, Evans's analysis forms the core of the argument here on bioethics as biopolitics. See Evans, *Playing God?*

36. Evans, *Playing God?*, 11. See Sinsheimer, "Prospect for Designed Genetic Change."

new eugenics as much as noting the blatant turn from such "thick" debates towards the "thin" analysis and language based in autonomy as a newly institutionalized value within bioethics.

Drawing from Weber, Evans further traces the establishment of bioethics's "thin" formal rationality—a rational *calculation* seeking the most efficacious ends towards "predetermined or assumed ends"—over a "thick" substantive rationality—one that asks if means of HGE are "*consistent with* ultimate ends or values."[37] The difference between these two rationalities is significant as "calculation" within "predetermined or assumed ends" follows a more efficient mode of instrumental rationality, whereas seeking "consistency" within ultimate ends remains open to transcendence beyond theoretical and political edifices. Within the structure of bioethics's predetermined ends, however, arguments and positions can be first "produced" then "selected" and then "institutionalized." Articles and books are produced and published *somewhere* within the field of bioethics before being selected "by the environment," an environment in ethics that includes people who are perceived as authorities both on judging HGE as well as in rendering tenure decisions for writers. Finally, after an argument has been to some degree sifted or evaluated by the field, it becomes "reified" as "common sense" or perhaps dubbed "factual," hence taking on an autonomous and authoritative "life independent of its creator":

> Within this perspective, the rise of formally rational forms of argumentation can be seen as the cumulative result of the actions of many individuals who create competing arguments to extract resources from their environments. Explaining in a cumulative sense why the supporters of formal rationality "won" in this competition and why the supporters of substantive rationality "lost" requires examining the factors that allowed some actors to disseminate their arguments, while others could not.[38]

What such a schema presents is a way to trace the influences on how bioethical arguments are acknowledged and valued, and hence why the language of bioethics is distinctly political. Evans identifies within the genre of bioethics itself a threefold division into foundational, clinical, and public debates. Foundational bioethics engages issues and topics related to "broader societal concerns" such as HGE, health-care reform, and transhumanism, while clinical ethics addresses the bedside dilemmas of clinical medical care. Public bioethical debate, however, moves beyond both the foundational-societal and clinical levels, as this is where the societal elites

37. Evans, *Playing God?*, 13 (emphasis original).

38. Evans, *Playing God?*, 26.

or professionals engage the foundational and clinical issues within their specialized domain. In principle, there is no real line between the public spheres and foundational or clinical, and yet ironically the way the public most often interfaces with bioethics is through the medium of specialized, social elites. Hence "the purpose of the public bioethical debate amongst professionals is to influence the beliefs and values of the public, to come to some modicum of consensus, or in some cases to represent public opinion to policy makers."[39] Naturally, within this framework, the question to ask is: *In what way* are specialized, academic bioethicists to function as a neutral medium and representative presence? Or, is it not more accurate to note that the bioethicists in question actually serve to screen and direct the debate towards their desired ends, whatever those may be? If we accept Evans's analysis of institutionalized, formal rationality, bioethics may be presented philosophically, implying profound moral deliberation, but its essential impulse and trajectory will be a politics motivated by power.

To rephrase this slightly, if we define a *bioethicist* as one who navigates primarily amongst the profession's form of argumentation, this entails that the bioethicist must learn the "principlist" *lingua franca*, and appeals to autonomy, beneficence, nonmaleficence, justice, and the like will henceforth constitute the center of all ensuing bioethics discourses. Existing within bioethics's principlist "system of argumentation" is, for Evans, what sets apart a professional bioethicist in contrast to philosophers, theologians, scientists, and/or physicians who interact with the issues of bioethics. Moreover, Evans suggests we ought to envision the bioethics profession as one amidst other competing professions within a "competitive ecological system." Here we can begin to see the raw calculations and power games that characterize the "thin" systems of institutionalized, formal rationality. The competition centers on who exactly has primary "jurisdiction" to influence significant policies such as regulating HGE, transnational pharmaceutical research, or perhaps mandating participation in human subjects research as a part of US citizenship, an argument Rosamond Rhodes first presented in 2005.[40]

Tracing such jurisdictional strands leads us to consider further the nature of power and the powerful. Beyond "mere" public opinion, the implications of this power include certificates of authority via state licensing for physicians, nurses, and healthcare practitioners, as well as institutional affiliation, access to grant funding, and more.[41] Moreover, following "market

39. Evans, *Playing God?*, 34.

40. Rhodes, "Rethinking of Research Ethics," 5, 7–28.

41. Evans, *Playing God?*, 28–32. Evans is not blind to shades of jurisdictional control when speaking of different domains such as medicine and ethics. Medicine typically seeks "full jurisdiction," where state-issued licensure regulates the profession,

research" and a "role delineation study" that ran from October 2016 to February 2017, the American Society for Bioethics and Humanities (ASBH) has launched Healthcare Ethics Consultation Certification (HCEC) as a "voluntary, professional certification program that will validate the professional knowledge and legitimacy of individual healthcare ethics consultants."[42] This certifying body has yet to define its criteria for ethics credentialing, but the jurisdictional terrain clearly is being marked and claimed. Here we are quite far away from the language of philosophers deliberating on the good or of priests offering words of pastoral counsel. The language of bioethics in this purview becomes a positional platform and agency of action—this language is a *techne* that grants authority amidst moral questions of *bios*:

> For example, the rising legitimacy of scientific thinking in society due to the efforts of scientists and others has worked to the benefit of medicine—whose form of argumentation presupposes science—and against folk healers of various types, whose arguments are not based on science. In the case of the HGE debate, in the late 1960s people who were not involved with the HGE debate successfully institutionalized a generalized questioning of authority by the public, and specifically the idea that scientists do not always work in the people's interest. The institutionalization of this idea among the general public, who were the empowered decision-makers of that era, led to a delegitimation of scientific authority (i.e., jurisdiction) in general—leaving scientists open to challenge in the peripheries of their jurisdiction (such as HGE).[43]

Evans's example of HGE is one amongst numerous cases that illustrate discourses, practices, and institutions within the workings of bioethics. Evans's analysis, moreover, traces a sociological pattern, and within this pattern the broad parallels to Foucault's biopolitics are hard to miss.[44]

---

whereas bioethicists have taken an "advisory jurisdiction" primarily within the domain of medicine and the sciences. Regardless, the transition from substantive to formal rationality yields a malleable body of reason that can be maneuvered amidst jurisdictions.

42. American Society for Bioethics and Humanities, "Certification for Health Care Ethics Consultants," para. 1.

43. Evans, *Playing God?*, 31.

44. Therese Lysaught notes that Evans's Webberian account supports biopolitical thinking but that Evans would need to engage Foucault directly to move specifically into biopolitics. Lysaught clarifies that Evans does not "attend to the obvious relationships . . . between the human genome project and the management/production of human bodies." Also, Evans does not note the important relationship between "the reconfiguration of bioethics and growth of the biotech industry." Whereas Evans critiques the reduction of the four principles in bioethics down to one primary end, autonomy,

Foucault's analysis and focus on the material reality of bodies within medicine ("biomedicine") are well known, particularly in *Birth of the Clinic* (1963), and bioethics is a natural consideration alongside biomedicine. One can see in Foucault a Kantian search for the conditions of possibility of knowledge, for to identify issues rightly in the present time, as Bishop notes, "we must first take a good look around at how the problems present themselves . . . [given that] solutions emerge in particular historical and political circumstances and are instantiated in practices."[45] In other words, our knowledge and practices arise from concrete historical realities and political spaces. Moving away from Kant, however, Foucault traces the practices and powers shaping medicine, psychiatry, and other human sciences not for a Kantian hope of "discovering" transcendent knowledge as much as discerning the social and cultural elements involved in "creating" knowledge. Hence Bishop points, for example, to the National Health Service that followed in the wake of Francis Galton's statistical medicine that was so influential in the nineteenth century:

> The modern idea of evidence-based medicine is an attempt to obtain statistically well-grounded reasons for acting, but it still remains, not a medicine of individuals (despite its claims), but a medicine of populations dispersed in space. Foucault eventually names these forces at work in society "biopolitics," in which life itself is ruled and conditioned, not only in the sense of a power that oppresses by naming and normalizing life, but also in the sense of creating the possibility of new vitalities and new freedoms for citizens.[46]

hence diluting the bioethics debates, Lysaught suggests that autonomy furthers the "internal logic of the profession of bioethics" to further the desired ends of the profession (Lysaught, "Docile Bodies," 404n19). Further, the point here is not to articulate an exhaustive Foucaultian analysis of bioethics but rather to consider biopolitics as a key touchstone for understanding the role of institutions and politics when analyzing the ethics of the body as *bios*.

45. Bishop, *Anticipatory Corpse*, 32.

46. Bishop, *Anticipatory Corpse*, 43. On the matter of power functioning as a possibly "oppressive" and "creative" force, Bishop and Jotterand note that Foucault's understanding of the role of power in biopolitics shifted from his earlier and later works. In his initial lectures on biopolitics, "Foucault articulates a notion of power contrary to that of his earlier works (*Madness and Civilization*, and *Birth of the Clinic*) where power is primarily a restraining and oppressing force. His new use of the notion of power in biopolitics takes on the character of the sustenance offered by politics to create structures that enable populations to achieve new freedoms and vitalities. The state comes to wield power over life, but now an enabling power that is not just repressive, creating instead the conditions for new capacities" (Bishop and Jotterand, "Bioethics as Biopolitics," 206).

In this vein, Lysaught argues for transnational research ethics as a compelling example of bioethics as biopolitics, and a gripping example of this can be seen in scandalous pharmaceutical trials conducted by Pfizer in war-torn Kano, Nigeria, in the late nineties. Lawsuits totaling nearly $7 billion (US) were brought by Nigerian families and government for major ethics violations, including withholding proven therapies for children, failing to inform families of the trial's risks, etc. What is remarkable here is how inept a "four principles" line of institutionalized bioethics thinking is to address an international case that includes conflicts of interest, massive corporate profits, and the humanitarian questions of international research on third-world children. To be sure, it *might* be helpful to discuss autonomy and degrees of informed consent in a non-Western environment where such terms may or may not remotely mean the same thing, but what matters more concretely is tracing the lines of power and control over bodies. Honesty and openness regarding the means of financial profit, influence, and health of populations, not *the* four principles of bioethics, are more fitting points of focus. Lysaught's central argument is that tragedies such as Pfizer in Nigeria cannot be adequately measured through the "standard narrative" of bioethics that focuses on a "rhetoric of freedom, public discussion of goods, protecting and advancing pluralism, and simply being about procedures rather than goods notwithstanding."[47] In place of such ambiguous, procedural language—a language that is hardly easily translated into Sub-Saharan West Africa—bioethics as biopolitics brings to light the various social strategies of policing and controlling populations that are wrought under the banner of furthering the goods of the social order.

In a striking irony, moreover, by clinging to the standard narrative of autonomy, freedom, and pluralism, bioethics can bring about harm, and this harm may do worse than muddy the waters via ambiguous principles. Bioethics itself can serve to veil the true intentions of institutions, governments, and others seeking "advantages for the general social good" at the expense of the weak and powerless who become "docile bodies" employed as needed for the betterment of (select) public health. Hence bioethics becomes an agency "to produce, organize, and manage the bodies of real, human persons—to police and control populations—toward the ends of larger institutional agents such as the state or, more recently, the biotech industry."[48] Lysaught contrasts this unpopular assessment with the "conventional wisdom" that describes bioethics as a protecting agency for the rights of persons in human subjects research. Following forty years (1932–72) of research atrocities—both

47. Lysaught, "Docile Bodies," 396.

48. Lysaught, "Docile Bodies," 387.

known and unknown at respective times—in Nazi Germany, Tuskegee, Guatemala, and more, *The Belmont Report* was produced in 1979 with the intent to preserve individuals' autonomy and rights of liberty. In the case of the Willowbrook State School scandal (ca. 1947–87), however, Saul Krugman (New York University), Robert McCollum (Yale University), and others intentionally infected scores of disabled and institutionalized children with hepatitis while seeking a hepatitis vaccine for the broader population. The justification for infecting these children came from the assumption that they would almost certainly come down with hepatitis anyway as a result of the poor conditions already present at Willowbrook. This atrocity can become mired and confused in theoretical, utilitarian discussions about *which* measure of social good ought to be employed when all the while the material reality of persecuted bodies is being ignored. Put more simply, academic deliberation can disguise human suffering.

Agreeing with Evans, Lysaught sharply challenges the standard birth narrative of bioethics growing "nobly" out of reflections on the Nazi Holocaust and other human subject-related research travesties. That is one part of the story, but what is missing is the storyline of bioethics emerging as a "mechanism for shaping and controlling the *hoi polloi* so that the growing Leviathan of biomedical research could quietly continue to pursue its own ends behind the scenes."[49] One of those ends turns out to be that bioethics has selected ethics research itself as a chief "good," and this means that in countries lacking clean water and other basic resources, international research from large corporations is often valued above bringing aid and relief to those in need,[50] or that sickly, institutionalized children who are "doomed already" are hence viable subjects for some of the most undesirable forms of research.[51] Within this assessment, it is easy to see how something such

49. Lysaught, "Docile Bodies," 393.

50. "Given the new political economy of globalization, one could argue that, although states remain proximate ends of disciplinary power, a new agent and end of disciplinary power are transnational corporate entities and that individual states now serve as intermediary agents of this [sic] power-mobilizing bodies in service of transnational social order" (Lysaught, "Docile Bodies," 397).

51. Naturally, this line of thinking can extend to research on embryos that can be ranked from nonviable to viable yet "unselected." If the assumption is that embryos are merely material cells that exist in a prepersonhood state, they are perfectly suited for research on whatever ends desired. This is not to equate embryonic research—before the "morally appointed" day 14 cutoff—with research at Willowbrook, but rather to point out the materialistic logic common in both instances. Few today see materialistic logic with embryos as problematic, but such materialistic logic in service of the greater social good as is dominant with embryonic research leads neatly to justifying Willowbrook, the Pfizer trials, or even to the Nazi holocaust. The only difference between these cases stems from where or when one decides that material life is meaningful. Other

as *The Belmont Report* becomes canonical within bioethics—a true source of Evans's formal rationality that is then further cemented as a touchstone in Beauchamp and Childress's *Principles of Bioethics.* All the while, what is glossed or lost entirely amidst the "rhetoric of freedom and autonomy" is the "surveillance and control" of individuals and populations as docile bodies by institutions and governments all participating in what amounts to nothing less than a new "transnational economic order."[52]

Additionally, within the political philosophy sphere, we may glimpse another, complementary critique of the common bioethics narrative by considering the role of deliberative democracy in bioethics. As Griffin Trotter has argued, bioethics is fundamentally centered on "cognitively higher-order things like rational consensus, human rights, and social justice," and yet there is more to be explored behind this cognitive domain.[53] Humans are not purely cognitive beings with rational access to absolute truth. Human beings are "moral, believing, narrating animals—as opposed to both rational, acquisitive, exchanging animals and genetically adaptive and governed animals."[54] Moreover, humans desire and seek after forms of power to gain dominance such that rationality itself may serve as the vehicle or foil for accomplishing what is believed or desired on a deeper level:

> Rationality is certainly an important weapon in our arsenal and, as with our other weapons, we experience pleasure in wielding it skillfully. For human beings, there are no pure instrumentalities. But rationality is always mixed with other affections such as anger, greed, pride, lust, and prejudice (to name a few)—and typically it is employed, at least in part, to serve them.[55]

Trotter points out that bioethicists possess a strong tendency to seek political influence as an expected means towards advancing their respective

---

questions of inherent meaning within research, or inherent value in certain types of knowledge (i.e., playing God via limitless knowledge of the body) are not given space within materialistic logic.

52. Lysaught, "Docile Bodies," 402.

53. Trotter, "Bioethics and Deliberative Democracy," 236. Trotter defines "deliberative democracy" as a democracy focused on "the legitimizing role of deliberation among citizens" as contrasted by a democracy emphasizing "aggregate" or "vote-centric" modes of democracy. Trotter draws this distinction from Kymlicka, *Contemporary Political Philosophy,* while arguing that "the convergence or consensus-oriented school has held sway [in bioethics] for well over a decade." The degree of consensus may vary amongst different parties, but the point to note here is the commitment of bioethics towards the "deliberative process and the goal of moral consensus" (Trotter, "Bioethics and Deliberative Democracy," 237–38).

54. Smith, *Moral, Believing Animals,* 118.

55. Trotter, "Bioethics and Deliberative Democracy," 236.

positions, and that "waxing eloquent" about rational consensus, human rights, and more disguises their political ambitions pursued through deliberative democracy. The focus on institutions and power struggles aligns well with Foucault, but Trotter directs our attention to Hobbes's fivefold critique of democratic deliberation. This critique and alignment with bioethics is worth summarizing briefly given the way that Hobbes's five theses align with bioethics as biopolitics:

First thesis: *Participatory democracy tends to devolve into aristocracy.* Because participatory or deliberative democracies allow free expression, the natural result is that "contenders and public personalities" rise to the surface, and ultimately the "most adroit manipulators of public sentiment" will hold power. Likewise, in bioethics, the influential experts gain political connections. For example, in the United States, the National Commission, the President's Commission and the National Bioethics Advisory Commission (NBAC) are notably filled with appointments from academic bioethics positions. Trotter states: "the ostensible rationale for these appointments seems to include the dubious supposition that bioethicists' extensive training in a diversity of philosophical systems was sufficient to ensure an objective, nonpartisan approach to the issues at hand—despite the fact that the appointees leaned uniformly leftwards towards the Democratic Party."[56] The notion of nonpartisan politics and bioethics are equally mythical, and this is advantageous to those who would enact the aristocracy.

Second thesis: *Participatory democracy begets unproductive disputes.* For Hobbes, much of the discourse within politics distracts one's focus from more meaningful activities of life with family and friends. "Profitless subjects" are those matters of disputation where individuals should simply "go their own ways." Numerous issues in bioethics fall within this concern. Where variant "fundamental moral beliefs" beget citizens with little to no moral common ground, what consensus should we expect on issues such as abortion, physician-assisted suicide, embryonic stem cell research, and more? Bioethics dialogue on these matters may influence legislation to varying degrees, but the central moral controversies remain. "For Hobbes, public advocacy of this nature is profitless, not merely because it cannot succeed in producing a robust moral consensus, but more importantly because liberty and individual felicity would be enhanced by allowing opposing individuals and groups to establish their own practices, free from government interference and from each other."[57]

56. Trotter, "Bioethics and Deliberative Democracy," 239, 238–39.

57. Trotter, "Bioethics and Deliberative Democracy," 240, 239–40.

Third thesis: *Participatory democracy begets excessive government power and dominion.* This thesis confronts the difficulty of balancing power and dominion. Trotter suggests that power may be understood as the "ability to enforce obedience" while dominion is the "scope of things enforced." The challenge Hobbes sees is that within a deliberative democracy where all participate in its rule, dominion is easily confused with power. As "the many" are involved in ruling through deliberation, all bring various interests and backgrounds, and political power becomes but another means towards securing one's own interests against those of others. Trotter quotes Hobbes's remark that in a popular democracy, "there may be as many Neros as there are orators who soothe the people," and a likely result of this is that political trade agreements are perpetually forged, seeking *quid pro quo* amidst committees, commissions, agencies, and more. An enormous "hyperplasia" of political power becomes the norm in place of a minimal system of laws existing only to provide basic conditions of peace. Not surprisingly, bioethics falls easily within this conflagrated power structure as bioethicists seek federal grant money focused largely on research projects aimed at regulations and protections run by the government itself. As such, "bioethics is largely in the business of enhancing the dominion of government, even as it enhances its own."[58]

Fourth thesis: *Participatory democracy breeds excessive collective aspirations.* Building upon the previous theses, the fourth thesis stresses the "suffocating profusion of collective thinking" that emerges when citizens "conjure grand visions"[59] of what might be achieved with political allies. Differences may be held lightly in favor of pacts forged on compromised positions, all with the idea of gaining some common shared ground. Yet, Hobbes argues that these collective aspirations can exceed human capacities, effective political control, as well as what is needed for peace. As a fourth critique, these collective aspirations can be based on irrational fears, which is an especially dangerous tactic of inciting irrational public hysteria.

These elements of excessive collective aspiration align strikingly with bioethics as a field strongly characterized by aims for universal, common morality. Hence Trotter points us to Beauchamp and Childress's claim that common morality is comprised of "core tenets in every acceptable particular morality that are not relative to cultures, groups, or individuals." Literally, "all persons living a moral life" know this moral code.[60] Within such a vision of morality, bioethics fundamentally exists as a project of public,

58. Trotter, "Bioethics and Deliberative Democracy," 241, 240–41.

59. Trotter, "Bioethics and Deliberative Democracy," 242.

60. Beauchamp and Childress, *Principles of Biomedical Ethics*, 3.

national, political morality. Hence we can see bioethics as a collective governmental campaign seeking: (1) "a degree of national moral consensus that is unworkable"; (2) "inordinate political influence over matters that cannot be controlled by national governments (for instance, elimination of health disparities)"; (3) to champion "projects that have little or nothing to do with securing peace"; (4) and to stoke "exaggerated public fears (for instance, by constantly portraying America's 45 million uninsured citizens as lower income individuals desperately in need of healthcare."[61]

Fifth thesis: *Rational consensus will never be the primary motive in public deliberation.* Hobbes's fifth thesis brings to attention political power as the driving motive within deliberative democracies. People act politically to gain political power, and this political power is "useful instrumentally as a means to living well,"[62] despite authentic, foundational differences. Moreover, for some, political compromises and pacts may stem negatively from fear of others. Equally then, we should wonder what most motivates bioethics? How do assertions of moral consensus from leading bioethicists differ from other political strategies?:

> Moral consensus is not a desirable pursuit, at least in the political realm, because political processes aimed at moral consensus exacerbate each of the aforementioned dangers of participatory democracy. Founded as it is on the false claim that extensive moral consensus is possible, the quest for consensus requires multiple layers of deception. Citizens must be deceived into believing that the opinions of statesmen or moral experts are morally authoritative for the whole group (Hobbes's sovereign would never pose such a ludicrous claim). They must be deceived into believing that their well-being is enhanced by acquiescing to the purported moral consensus. And they must be deceived into believing that this acquiescence is in important respects autonomous and voluntary (despite the obvious fact that it is coerced).[63]

In sum, the language of bioethical consensus has less to do with moral consensus than it does with the reality of political power and processes. The language of bioethics can and does function significantly as a moral *techne* within a greater system of preselected and predetermined ends. The scope of bioethics language as moral *techne* naturally can shift from blatant examples of abuse such as Willowbrook and the Pfizer trials in Nigeria to

61. Trotter, "Bioethics and Deliberative Democracy," 242, 241–42.

62. Trotter, "Bioethics and Deliberative Democracy," 242.

63. Trotter, "Bioethics and Deliberative Democracy," 243, 243–44.

more veiled national democratic processes of manipulation evidenced in the NBAC. Yet even more can be discerned under the heading of biopolitics by looking further into a topic mentioned earlier: the integral role of theologians in the early days of bioethics.

3. Shifting Language Games: From Theology to Bioethics

An ethicist is a former theologian, who does not have the professional credentials of a moral philosopher. —James Gustafson[64]

With a few exceptions, the theologians and the philosophers, who were partners in the early medical ethics, merged into one—the bioethicist. —Albert Jonsen[65]

The primary contribution of theologians to the new field of bioethics, for Jonsen, was to bring "a sharper concept of ethics" stemming from the "classical discipline" of theology.[66] In this way he rather simply compares theologians to philosophers, both of whom began to address the "new medicine" of the twentieth century from their respective disciplines.[67] Jonsen's narrative unfolds with the textbook-styled, objective voice common to academics, and yet even Jonsen's neutral tone cannot ignore the twentieth-century cultural instability affecting theology and bioethics; Roman Catholic moral theology did not address bioethics from a place of traditional theological security. The *aggiornamento* that marked Pope John XXIII's papacy and the innovations of Vatican II (1962–65) represent a significant reexamination of traditional formulations affecting the whole of the Roman Catholic tradition and many surrounding cultures influenced by Roman Catholicism. Within this context, Roman Catholic theologians in bioethics encountered the "turmoil" and "inner debates about methods" and "dissensions" regarding central doctrines, particularly as issues of masturbation, abortion, contraception, in vitro fertilization, euthanasia, and more gained attention. Jonsen particularly mentions Richard McCormick (1922–2000), Charles Curran (1934–), and Bernard Häring (1912–98) as Roman Catholic theologians who sought to nuance and challenge "pronouncements of the

64. Gustafson, "Theology Confronts Technology," 386.

65. Jonsen, *Birth of Bioethics*, 58.

66. In this section, I use the term "theologian" primarily in the sense of an academic discipline and not in the holistic sense often summarized by Evagrius Ponticus's "if you pray truly, you are a theologian" (Evagrius, "On Prayer," 1:62).

67. Jonsen, *Birth of Bioethics*, 34–64.

magisterium" on issues newly identified under bioethics.[68] Similarly, Protestant theologians were hardly removed from such shifting methodologies and allegiances, particularly in the wake of the early-nineteenth-century Social Gospel movement and the 1960s ecumenical movement between protestants and Roman Catholics.

In particular, Jonsen highlights a "trinity" of theologians: Joseph Fletcher (1905–91), an Episcopal minister who later would reject Christianity;[69] Richard McCormick, a Jesuit moral theologian; and Paul Ramsey (1913–88), a Methodist professor. Few would disagree that these theologians were powerful figures amidst the 1970s "golden age" of bioethics, and more will be said particularly of Ramsey shortly, but first it is notable how Jonsen speaks of the "strange happening" for Christian ethicists "on their way to bioethics."[70] Namely, denominational identities and robust theological discourses "faded" amidst the migration to bioethics. James Gustafson and Stanley Hauerwas were among the first to point out the significance and impact of this change in identity and language. Gustafson wrote in 1978 ("Theology Confronts Technology and the Life Sciences") that the growing interest between theology and technology and the life sciences was clearly evident, and yet the actual theological interaction between these spheres was highly suspect. Many of these "ethicists" were theologically trained, and yet their explicit appeals to a particular theological or religious "outlook" were thin and "opaque." Gustafson commends Paul Ramsey as an exception to this trend in that he ardently described himself as a "Christian ethicist" who did not write with "some hypothetical common denominator," and yet Gustafson's student Hauerwas is less convinced that Ramsey's "execution matched his candor."[71] Hauerwas jests that all of the theology in Ramsey's *The Patient as Person* (1970) is in the

68. Jonsen, *Birth of Bioethics*, 37.

69. Fletcher eventually exchanged his post in a divinity school for one in a medical school, and he made explicit his turn to secular ethics by declaring himself an act-utilitarian. See Jonsen, *Birth of Bioethics*, 42–47.

70. Jonsen, *Birth of Bioethics*, 57. As further evidence of the role of theologians early in bioethics, we can note that theologians such as Paul Ramsey, James Gustafson, Richard McCormick, and Karen Lebacqz served on national policy groups such as The National Commission on the Protection of Human Subjects of Biomedical and Behavioral Research (1974) and the President's Commission for the Study of Ethical Problems in Medicine and Biomedical and Behavioral Research (1979). Additionally, as noted earlier, academics with theological training were influential in the launch of organizations such as The Institute of Religion at the Texas Medical Center in Houston (1954); the Institute of Society, Ethics, and the Life Sciences (1961; now named The Hastings Center); and the Kennedy Institute of Ethics at Georgetown University (1971).

71. Hauerwas, "How Christian Ethics Became Medical Ethics," 15.

preface, and yet the concern is less with the "amount" of Ramsey's theology than it is with the kind of theology being presented, a theology methodologically enacting the social gospel movement:

> But why should [Ramsey's] understanding of the ethics of the Gospel be called protestant liberalism? Quite simply, it allowed Ramsey to think that the nicer issues of theology, such as trinitarian and ecclesiological issues, were largely tangential to ethics . . . I am not accusing Ramsey of the kind of reduction so characteristic of much of Protestant liberal theology, e.g., Christ means love of the neighbor. Rather his position, in a sophisticated way to be sure, accepts the presumption that the gospel has a moral upshot. The Christian essence can therefore be known without the frills.[72]

While Ramsey states his firm commitment to being a "Christian ethicist," his ethics may or may not appeal explicitly to a Christian vision of the good, and in this way, Ramsey's ethic can be seen as deontological.[73] In a 1990 letter to Hauerwas regarding the charge of "inadequate theological warrant," Ramsey maintains that his ethics shares less with "the humanists'" commitments and rather stems from a commitment to the "giftedness" of life. Hence Ramsey sees his foundational commitments—even when theologically tacit—as on par with those of Hauerwas, intimating a difference in style, not substance. In pursuing an ethics translated from a Christian into a neutral discourse, Ramsey maintains "that Christian special ethics would still come to the conclusions I do."[74] Such ardor aside, Ramsey's assumption that Christian ethics and natural reason will align remains suspect. Hence agreeing with Hauerwas, Engelhardt identifies Ramsey as part of the early Christian bioethics presence that was "amphibious" in maintaining a Christian label without "affirming any essentially Christian understandings."[75] This was a bioethics that "had insufficient resources to show how Christian practice might make a difference for understanding or forming the practice of medicine," and this leaves Hauerwas to ask: How was Jesus' preaching of the kingdom at all needed for "such an ethic?"[76]

72. Hauerwas, "How Christian Ethics Became Medical Ethics," 22–23.

73. "I am aware that Ramsey's position is more complex than the description deontological can comprehend, but I find it hard to see how Ramsey takes us beyond what John Milbank has identified as deontological liberalism" (Hauerwas, "How Christian Ethics Became Medical Ethics," 25).

74. Hauerwas, "How Christian Ethics Became Medical Ethics," 18.

75. Engelhardt, "Recent History," 147; 146–50.

76. Hauerwas, "How Christian Ethics Became Medical Ethics,"16, 24–25.

In what sense can we identify Ramsey's ethic as a Christian approach to medical ethics? Is his commitment more to a neutral rationality or to the Christian message? Or, perhaps it is best to identify his work as that of a believing Christian who employed a moral discourse that could be understood and appropriated by others without requiring any Christian belief on their part.[77] What emerges from Ramsey is an ambiguous and debatable approach regarding what sort of discourse Christian theology should adopt when addressing bioethical issues within the public square. Without a particular grounding or more concrete tradition of religious discourse, Ramsey remains an easy target for assimilation within the power structures and thin, institutionalized ends of bioethics.

The case of Fletcher's commitment to a firm theological discourse is easier to answer. Fletcher's legacy is that of a situational ethicist who after twenty-six years as a professor of social ethics at Episcopal Theological School in Boston explicitly exchanged his role as priest and theologian to become a secular medical ethicist. Whereas his early reflections on bioethics, such as *Morals and Medicine* (1960), were situated within a theological context, the conclusions of his moral thinking remained significantly at odds with traditional Christianity.[78] Fletcher "eschewed any prohibitive ethical claims that might halt or hinder research or treatment options, often embracing moral positions that most traditional Christian theologians would reject."[79] As Evans summarizes, Fletcher "begins the [1970s] era as a theologian and ends up a bioethicist,"[80] and Kass decried Fletcher as a progressive "theologian-turned-technocrat [who] . . . with the dead God as his co-pilot, is set to fly off into the wild blue yonder of limitless self-modification."[81]

In the case of Richard McCormick, we encounter a Jesuit ethicist entrenched within the natural law tradition, and yet McCormick frequently sought to reframe traditional Catholic natural law positions in a more liberal or pluralistic way to harmonize with a democratic society. Jonsen refers to McCormick's "classicist mentality . . . [of which he remained] conscious of both its strengths and weaknesses—and the need to correct or modify the

77. Engelhardt, "Recent History," 146.

78. In 1968, Fletcher appealed to a Kantian mode of personhood in addressing the desirability of euthanizing Down's syndrome babies: "[There should be no guilt] about putting a Down's syndrome baby away, whether it's 'put away' in the sense of hidden in a sanitarium or in a more responsible lethal sense. It is sad; yes. Dreadful. But it carries no guilt. True guilt arises only from an offense against a person, and a Down's baby is not a person" (Bard and Fletcher, "Right to Die," 59–64).

79. Bishop, "False Gods," 303.

80. Evans, *Playing God?*, 241n88.

81. Kass, "New Genetics," 60.

latter."[82] McCormick sought to be a moderate and a committed Catholic theologian, and he was explicit that "an ethic that claims to be theological will root itself in God—God's actions and purposes." Furthermore, "a Christian theological ethic is founded on the fact that something has been done to and for us, and that something is Jesus."[83] For McCormick's moderate, middle-ground stance, the main question turns on how he follows natural law as a way to defend positions on strictly philosophical grounds apart from doctrine. Whether or not such a nondogmatic stance is consistent or logically feasible is a matter particularly relevant to Christian bioethics,[84] and yet without engaging this question now, McCormick's legacy remains a player furthering the transition from theologian to bioethicist.

Even from within Jonsen's noncritical stance, the transition to bioethics was "a strange thing" for these Christian ethicists who were "educated in theological traditions, and sometimes clergy in their denominations." Even so, Jonsen casts aside this oddity by asserting "there were good reasons for this fall from faith." In place of historic, religious communities with thick, particular traditions and doctrinal commitments—commitments in modernity seen as "divisive" given that such "ruminations about human nature and destiny" call for "transcendent references of theology"—Jonsen instead imagines a neutral space for "the practical problems of bioethics" that can be "carried out quite satisfactorily in more mundane terms."[85] Within this modern purview, the theological language of early bioethics can—and even should—be abandoned for a bioethical analysis outside traditions of faith, and yet this thin, presumptive analysis entirely misses the enormous import of such a change in discourse. Altering language or language systems is not an insignificant, neutral, or mild change. As Evans staunchly argues, the very strength of theological discourse is to examine

82. Jonsen, *Birth of Bioethics*, 52, 52–55.

83. McCormick, "Theology and Bioethics," 5–6.

84. The question of natural law discourses in bioethics—specifically if natural law may be rightly understood within the Christian tradition—is addressed in detail in Engelhardt's *Foundations of Christian Bioethics* (2000). It is also worth noting Sulmasy's argument that "proponents of natural law like McCormick and Ramsey" were eventually dismissed in the face of "a progressively secularizing bioethics" that determined natural law as an "inherently religious (specifically, Christian)" method that could not accord with the official secular methodology of bioethics (Sulmasy, "In Defense of the Amphibians," 190). Such a reading of the history of bioethics may be persuasive to some, but the question remains as to the grounding of natural law itself. If, as Charles Curran argues, "the natural law by its very nature does not depend primarily on revelation and God's word, but on human reason," this internally seems a relatively compatible form of theorizing within the secular (Curran, "Curran," para. 4).

85. Jonsen, *Birth of Bioethics*, 57–58.

explicitly the "ends" and *telos* of things, whereas secular science subverts the question of ends amidst its naïve enthusiasm for progressive, technological means. Moreover, as these theologians-turned-bioethicists exemplify, the translation of rich, theological ends translated into mundane, secular language is entirely noncommensurable:

> The richness of the theological end could generally survive translation to secular language because, at [an early] point in the debate, any end that could be expressed in English could be used. Commensuration of theological ends, which would occur later on, involves the throwing away of information that does not fit into predetermined ends. To use a spatial metaphor, the theologians could create any (theological) shape they wanted and put it through a (secular) hole of the same shape. Later, theological square shapes would of necessity be stuffed into (secular) circular holes, losing part of their form as they passed through. These secularly stated ends were not assumed to be held by the population; the point was to convince the population to adopt them.[86]

In short, theologians at the birth of bioethics utilized theology where helpful, but where other language systems better fit the secular stated ends theology was jettisoned as cumbersome and irrelevant. Engelhardt is one who ardently decries this theologically "neutral" language of bioethics, and yet he raises the stakes on the question by addressing not simply theological bioethics but more particularly *Christian* bioethics. For Engelhardt, not just Fletcher but all three of Jonsen's "trinity of theologians" became "marginalized" amidst the "secularization of the public fora and public spaces of the West." For Engelhardt, the deeper critique is that these Christian ethicists turned bioethicists "due to a form of self-demolition tied to a failure to be authentically Christian."[87] Hence the shift from theology to bioethics is certainly related to the cultural shifts and context of secularism, and yet particular individuals and voices speaking in Christian bioethics in the 1970s can and should be challenged on the grounds of what it means to speak *as Christians*. Engelhardt's critique aligns closest with Hauerwas as Hauerwas distinguished his Christian message from principlist discourses:

86. Evans, *Playing God?*, 65–66. Evans further points to the particular proclivity of Protestant theologians to differ and change amidst their debated theological ends. In short, Protestantism tends to follow a methodology of appropriating disparate biblical themes, and this then enables multiple theological debates around different ends.

87. Engelhardt, "Recent History," 147.

> For by speaking from a theological perspective I do not pretend to speak from principles that are or should be shared by everyone in our society. You should also know that my own methodological presuppositions in this respect are not widely shared among those that work in theological ethics. Rather the assumption is that theological ethics must develop arguments that should compel consent from all rational subjects irrespective of their religious convictions or lack of religious convictions. Of course, that results in the somewhat ironical state of affairs that committees such as this one invite representatives of religious communities to show how their communities' particular convictions throw light on an issue only to be told that Christian views on the subject are not necessarily related to their religious convictions. Christian ethicists therefore say what any right thinking moral philosopher or person would say. Well I simply do not believe that. It will be the heart of my argument that theological beliefs do make a difference for how *in vitro* fertilization is understood.[88]

The import of this, ethically, is that traditional religious moral convictions do not equate methodologically or ontologically to secularist or democratic principles. Much can and will be lost in translation, and this is not a call to despair for all communication but rather to point out the need for a closer understanding of the ontological and theological realities that must be acknowledged. The point in challenging Fletcher, Ramsey, and McCormick's mainstream bioethics language is to note the vast difference between theorizing about the "right or wrong" of some biotechnological possibility and determining who we are *to be* before God. The thickness of Christian ends places a different *ordo* on ethics that directs us to questions of being and ontology—of what and who we are—as an essential step before seeking the relative goodness of an action or biotechnology. In this vein, adding to Hauerwas and Engelhardt's critiques of theologians-turned-bioethicists, Bishop reminds us that such an ontological viewpoint is necessary to grasp the "metaphysical moral claims implicit in medical technology" that rival the Christian ontological worldview.

> The medical establishment insisted that theology thin out its language before they were permitted at the table; and they did so precisely because these voices threatened the implicit ontologies and teleologies of medical science. Put differently, the ontology and teleology of medical science are not thin. It is in fact robust; however, it is couched in a very different sort of language.[89]

88. Hauerwas, "Theological Reflections," 2–3 (emphasis original).

89. Bishop, "False Gods," 306.

In other words, Bishop furthers the biopolitics of Engelhardt and Evans, arguing that the transition of "theological ends" into "thin secular ends" was not merely a conforming to "formal reason" but actually was a robust clash of the ontologies and teleologies between secular medical science and the Christian tradition. Moreover, as Hauerwas notes, the motivations for shifting from theology and Christian particularity are not difficult to imagine, and these motivations are the biopolitical elements that should be unpacked within the narrative of rival ontologies:

> So medical ethicists, being the good priests they are, went to where the power is in liberal societies—medical schools. Kings and princes once surrounded themselves with priests for legitimation. Likewise, politicians today surround themselves with good social scientists to give those they rule the impression that they really know what is going on and can plan accordingly. Physicians, in an increasingly secular society, surround themselves with medical ethicists. God no longer exists, the sacred universe of values has replaced God, and allegedly ethicists think about values and decisions that involve values.[90]

These words from Hauerwas may seem harsh—for Ramsey and McCormack, if not for Fletcher—and yet Hauerwas's primary aim seems less a personal accusation and more an identification of the role of power that will serve as a default framework when unwittingly navigating from one robust ontology (the Christian tradition) into another (the secular, scientific medical system). Hence Hauerwas rightly critiques the simplistic narrative that bioethics in the 1970s was strongly inclusive of Christian theology and that this early inclusion of a theological discourse was largely phased out after Fletcher, McCormack, and Ramsey. Such a progressivist narrative misses the biopolitical and cultural reasons why theology was displaced amidst a growing secularism and the dynamics of America's morphing claims to be a "Christian nation."[91]

In addition to the narrative and critiques of theologians-turned-bioethicists, another helpful biopolitical viewpoint may be considered by

90. Hauerwas, "How Christian Ethics Became Medical Ethics," 14.

91. "Ramsey and McCormack surely took their Christianity as having been heuristic for them. However, given the seduction of the time, they did not place their reflections on bioethics within a distinctively Christian discourse. Instead, they tried to place Christianity and Christian bioethics within an idiom they hoped would be open to all. They did not invoke insights open only to Christians, for that would have appeared too sectarian and countercultural. Given America's having renounced its previous de facto if not de jure establishment of Christianity, it would have appeared un-American" (Engelhardt, "Recent History," 151).

examining the history of the phrase "sanctity of life." Here it is instructive to follow David Jones's assessment of the "sanctity of life" as a frequently employed "principle" that differentiates from a purportedly rival "quality of life" position. The "sanctity of life" position is typically associated with conservative or traditional ethics, and for some the "sanctity of life principle clearly has religious origins."[92] On the contrary, however, Jones draws from Leon Kass and others to point out that "sanctity of life" does not occur either in the Hebrew Bible or in the New Testament. Within the Roman Catholic document on assisted reproduction, *Donum Vitae* (1987), there is mention that "human life is sacred because . . . it involves the creative action of God,"[93] and yet Jones nuances the way *sacra* is employed rather than *sanctitas*:

> It may be noted that while *Donum Vitae* and John Paul II refer to life as 'sacred' (*sacra*), they do not speak of the 'sanctity' (*sanctitas*) of life and it may also be noted that even in these very recent documents the predominant language is not one of sacredness but of 'respect,' 'inviolability' and the 'right to life.'[94]

The use of *sacra* seems not to indicate a *principle* against killing or some inherent, essential nature of life itself possessing sanctity (*sanctitas*). Rather, *sacra* indicates a quality of life within some other context. A life should not be wrongfully taken as life is sacred (*sacra*), but the further move to proclaim an essential sanctity of life *qua* life is a different claim:

> The recent provenance of this phrase has been demonstrated by George Khushf (1996) after an exhaustive literature survey covering 4,800 libraries including libraries of congress in 26 countries. He was only able to find two significant references to the use of this phrase in its modern sense prior to the twentieth century, and "nearly all citations" were from after 1948. A distinct but related attempt by Andi Sullivan to find journal articles using the terms "sanctity of (human) life," "sacredness of (human) life" and "human dignity" also found no references before 1893 and only 239 citations before 1939 in comparison to 8,475 from 1940 to the present. Indeed, the great majority of references (6,467) were from 1970 onwards.[95]

92. Jones includes Keyserlingk, Otlowski, Kuhse, and Baytertz as examples of those who directly assume that sanctity of life "obviously" stems from a religious provenance based on the "vocabulary" alone (Jones, "Unholy Mess," 186).

93. The Catholic Church, *Donum Vitae*, para. 22.

94. Jones, "Unholy Mess," 187.

95. Jones, "Unholy Mess," 187.

Jones is clear that exceptions do surface. It is possible to find quotations regarding the sacredness of life, and yet the significant point is how rare these exceptions are. Moreover, the argument here does not entail that theological accounts of the sanctity of human life against wrongful killing should not be made. Rather, it is instructive to note

> that neither the fathers of the Church of Latin West or Greek East nor the scholastic theologians of the High Middle Ages did in fact develop such an account. They did not relate the Commandment against killing human beings to the sacredness of life. Understood as the name of a well-established "principle," "the sanctity of (human) life" is virtually an invention of the late twentieth century.[96]

So then, if not from within the Christian tradition, from where does the popular "sacredness of human life" principle arise? According to Jones and John Keown, it was Glanville William who popularized the phrase in his monograph *The Sanctity of Life and the Criminal Law* (1957), and it is pertinent to note that Williams does not preface his use of "the sanctity of life" by appealing to any previous citations, nor does he define the phrase.[97] Instead, the phrase is employed as a way to suggest that killing the innocent is wrong on the grounds of a "feeling" or "general opinion" stemming from a "*religious* heritage." Williams writes that "even the modern infidel tends . . . to regard all human life as sacred," and that this "feeling" is a "legacy of their religious heritage" and a profession within a society that "is likely to be coloured by that religion." Moreover, he notes that "this general opinion, whether religious or secular" might justify the punishment of a mother who kills her "monster or idiot child."[98] Clearly, for Williams, the prohibition on killing such infants stems—as a purportedly weak argument—from a religious legacy in the West, and hence Jones rightly aligns Williams with Peter Singer and others who focus on nonvoluntary euthanasia of intellectually disabled children as "idiots" or "monsters." What is particularly notable in this move to defame "sanctity of life" as a "religious" thing is how this negative conception of religion aligns so neatly within the imagination of the secular. As Milbank has noted, when the secular is accepted, an ensuing "desacralization" is to be expected as it is specifically not religion but rather the

96. Jones, "Unholy Mess," 189.

97. Keown and Jones, "Surveying the Foundations of Medical Law," 85–126.

98. Glanville Williams (*The Sanctity of Life and the Criminal Law*, 1957) as quoted in Jones, "Unholy Mess," 190 (emphasis original).

cosmos itself—including, inevitably, "the political, the social, the economic, the artistic—the human 'itself'"—that is the source of "sacral allure."[99]

Ironically, as Williams's "the sanctity of life" gained traction, the phrase became further cemented as some conservatives sought to counter Williams.[100] Hence attempting to contest Williams, they succumbed to the effects of his language game as the import of his phrase remained subtly derogatory. The sanctity of life phrase and pseudoprinciple arose from a secular source and remains ensconced in ambiguity given the lack of history, context, or definition for the phrase. Without a *locus classicus*, the possibilities for clear communication with the phrase require one to stipulate a new definition, and the primary connotation is dismissive. Jones explains:

> The connotations of this language are part of a deliberate attempt to distract from fundamental issues of justice, solidarity, and human rights and falsely to imply that the legal protection which is due to vulnerable human beings is based only on religious sentiment . . . A direct consequence of the recent emergence of this language, and of its origin as a description of an approach that is being rejected, is that there is no agreement as to the definition of this "principle." . . the description is external and alien to the intellectual tradition being criticized.[101]

In short, without any substantive definition for "sanctity of life" that actually matches the nuanced and layered reasoning of those like traditional Christians who are concerned with wrongful killing, the phrase exists primarily as a straw-man fallacy for the sake of marginalizing the other position. "Sanctity of life" is then inherently reductionistic when employed to argue for the worth of mere physical existence. Such vitalism allows Fletcher, Singer, and others to then maintain that there is more value in "life" beyond biological existence, and hence it becomes very reasonable, even merciful, to practice nonvoluntary euthanasia for "idiot" children with Down syndrome who are not deemed full persons. Personhood is a value that exists beyond merely being alive, and hence children with Down syndrome do not possess the requisite "quality of life" needed to be persons. Within this reduced debate, the clash between "sanctity" and "quality" becomes a territorial dispute between antiquated, religious bias and modern,

99. Milbank, *Theology and Social Theory*, 9.

100. "It was only after Williams's book, and in opposition to it, that some prominent religious thinkers [including, amongst others, Norman St. John-Stevas (1964), Paul Ramsey (1968) and Daniel Callahan (1968)] took up the language of the 'sanctity of life' with a view to defending this 'principle,'" (Jones, "Unholy Mess," 190).

101. Jones, "Unholy Mess," 191.

secular, reasoned enlightenment. As expected, the winner of such a stacked debate is no surprise.

Moreover, Jones rightly notes the background philosophical assumptions—assumptions arising in twentieth-century moral theories—behind this reductionist schema as moral philosophy is limited to a dichotomy between deontological or teleological viewpoints. In this case, deontological thinking is associated with Kantian duties and rules, while teleological ethics does not mean Aristotelian final causality but rather final utilitarian calculations in the wake of Bentham and Mill. If and when this simple dichotomy is posited as the bioethical status quo, Fletcher is furthered in positing "sanctity of life" ethics as the deontological mode where (outdated) religious rules and vitalistic principles are rightly abandoned in favor of teleological theories that take into consideration "goods, needs, or values" of actual persons. The needs both of the "idiot" child with Down syndrome and the conflicted parents align in favor of nonvoluntary euthanasia. Quality of life rationally and theoretically trumps sanctity of life, and yet this is an ethical paradigm limited only within the immanent logic of secular modernity. Jones points to Aristotle and Aquinas as two immense opponents to this dichotomy of modernist logic, and yet ironically when "virtue theory" is added as a simplified third option amidst possible moral theories, we subsequently reduce virtue into yet another theoretical choice, yielding an oversimplification that moves from two to three ethical systems "options." The same can be said for ethics of care, feminist ethics, etc., and then—most damning of all—we see various religious systems added to the choices of "theories" as more and more options are added to the modern ethical playing fields.

The point for now is to note the language systems that inevitably emerge amidst the landscape of bioethics-as-biopolitics when secularism is the assumed common ground for ethics. The presumption that language or philosophy can be neutral practices within the secular falters before the critiques of biopolitics. In light of this, Allen Buchanan calls for an honest evaluation of bioethicists from the standpoint of social moral epistemology. Looking back upon the eugenics movements of the late nineteenth and mid-twentieth centuries, Buchanan notes the limited "conventional 'ethical autopsy'" that views the chief error of eugenicists as being too willing to ignore individual "rights" for the broader social "good." The correction for this presumably now-blatant error is "more Kant, less Bentham!" Buchanan rightly critiques this limited moral spectrum and turns to the biopolitical question of the expertise of a bioethicist.

> The distinction between rights-based and consequentialist moral theories and principles does little to illuminate such a

> subversion of conventional morality [when eugenics was naïvely, socially accepted]. We need to understand why so many well-educated laypeople, across the political spectrum, trusted the wrong "experts." In social moral epistemological terms, we must explain the phenomenon of *unwarranted epistemic deference*—showing how it came to be that certain individuals managed to present themselves to the public as scientific experts on human heredity and why the public was ready to accept what they said . . . We need an account of the epistemic *vices* of the supposed experts.[102]

So then, the question is not *if there are* but *what are* the epistemic vices of bioethics? What is the expertise of a bioethicist? Is the bioethicist a philosopher? An ethicist? At one point, bioethicists were largely theologians. Now, in what sense is bioethics a distinct discipline or profession? Is the more fitting title for a bioethicist actually a biopolitician? As Carl Elliott has pointed out, what is new is "ethics as a position of bureaucratic authority staffed by trained professionals." Ethics has been professionalized, which is to say that it has gained a corporate, bureaucratic structure, and while some may contend that this profession offers a service of critical thinking skills and certain expertise with regulatory policies, Elliott chillingly reminds us: "this position of authority . . . has been given, usually by virtue of their place in a particular bureaucracy." We live in an "age of expertise" empowered by the "virtue of efficiency and a claim to truth," and bioethicists stand as such "experts" who are "acculturated into a standard view of the world . . . [and] it is because of their power, not just their credentials, that experts are so difficult to challenge."[103]

4. Tar-Baby Principlism: The Pragmatism of Shifty Language

> *A picture held us captive. And we could not get outside it, for it lay in our language and language seemed to repeat it to us inexorably.*
> —*PI*, 115

To conclude this chapter, it is apt to reflect briefly on the language of principlism that has become more professionally enfranchised than any other in bioethics. Beauchamp and Childress's *Principles of Biomedical Ethics* (1977; 2019) has achieved eight editions (and counting?) and the status for many

102. Buchanan, "Social Moral Epistemology and Bioethicists," 292, 288–96 (emphasis original).

103. Elliott, "Tyranny of Expertise," 45–46, 43–46.

as the bible of bioethics. The "four principles" or "principlism" approach advanced in this text is nearly canonical within the field. Chambers observes that the *Principles of Biomedical Ethics* exhibits "a cool detached writing style that is expressed by an impersonal narrator" and is such a classic in the field that it is "often referred to not by title but by its authors: 'Do you have a copy of Beauchamp and Childress?'"[104] Moreover, to understand this popularity, we must uncover Beauchamp and Childress's (B&C) articulation of "common" or "universal" morality and to observe how explicitly they position their moral analysis within a flexible social and historical framework, a framework that begs for a biopolitical analysis. While B&C maintain on the one hand that the nature of *the* common morality is "not relative to cultures, groups, or individuals"—and they list examples such as "do not kill; do not cause pain or suffering to others; prevent evil or harm from occurring; rescue persons in danger; tell the truth; obey just laws, etc."—they simultaneously admit that common morality is not "ahistorical or a priori" but rather is "a product of human experience and history."[105] For B&C, this is a proper clarification and nuance on what exactly is universal or common: namely, the commonality in morality is something that arises "as a product" of a shifting and historically bound cosmopolitanism, and in their view the human experience of morality remains the same regardless of the shifting social and historical features that may arise. As Arras summarizes: "moral normativity is established historically and pragmatically through the success of these norms at all times and places in advancing the cause of human flourishing. Their account is thus historicist, but unlike most historicisms it does not embrace moral relativism."[106]

This formula for claiming universality and historicity in B&C is furthered by their distinction between thin "common" morality and a more concrete or "particular" morality. Particular morality extends from common morality and is not necessarily "shared by all cultures, groups, and individuals." Particular moralities may be Jewish moral norms located in the Talmudic tradition, or even professional moralities maintained, for example, in certain physician codes, and yet these particular moralities stem from what is common in morality.[107] Moreover, the balancing act between these two spheres presents B&C with a tension and space in which they can (and do) appeal to contradictory poles within the same formula; hence appeals to particular or common morality equally may be used to address moral tensions in either

104. Chambers, "Virtue of Attacking the Bioethicist," 286, 281–87.

105. Beauchamp and Childress, *Principles of Biomedical Ethics*, 3–4.

106. Arras, "Hedgehog and the Borg," 13.

107. Beauchamp and Childress, *Principles of Biomedical Ethics*, 5–6.

direction. Here B&C turn to John Rawls's "reflective equilibrium" as a type of coherentism that starts with "a body of beliefs . . . initially accepted without argumentative support" and balances this assumed foundation with more concrete, practical instances. The goal here is "equilibrium" between specific beliefs of an individual or community within the broader commitments of universal, common morality. For instance, when balancing organ donation (1) according to "the expected number of years of survival" with (2) a waiting list that grants equal opportunity to all individuals, B&C advocate for a "pruning and adjusting of beliefs" that accounts for moral norms behind both of these needs. An ideal state of reflective equilibrium is not possible, B&C posit, and hence they argue for a limited goal of "trimming, repairing, and reshaping" beliefs in a realistic but not "utopian" process focused on "particular moralities" that are "continuously under development, not finished products."[108]

Similarly, in the well-worn case of a Jehovah's Witness couple who refuses a medically needed blood transfusion for their young child, the clash between individual, religiously motivated autonomy, and the physician's (or society's) beneficence is clearly evident. Typically, ethicists approach this case as though the religious rationale is "out of the norm" and various strategies leading to a court order are offered to override the Jehovah's Witness couple's autonomy. Regardless how we judge the Jehovah's Witness position on blood transfusion, what is interesting is the inevitably "majority appeal position" that common morality employs. As Lee notes, "what ethicists seem to be doing under the banner of common morality is evaluating or judging the minority culture of Jehovah's Witnesses from the perspective of the mainline culture in the West (e.g., liberal individualism, humanism and mainline Christian denominations)."[109] In a passage that has been edited and reedited in each of B&C's 5th, 6th, 7th, and 8th editions, they address this matter through a middle ground:

> When moral disagreements arise, a moral agent can—and usually should—defend his or her decision without disparaging or reproaching others who reach different decisions. Recognition of legitimate diversity—by contrast to moral violations that warrant criticism—is vital in the evaluation of the actions of others. One person's conscientious assessment of his or her obligations may differ from another's when they confront the same moral problem, and both evaluations may be rooted in the common morality. Similarly, what one institution or government

108. Beauchamp and Childress, *Principles of Biomedical Ethics*, 441–42, 425–58.
109. Lee, "Problem of 'Thick in Status, Thin in Content,'" 526.

> determines it should do may differ from what another institution or government determines it should do. In such cases, we can assess one position as morally preferable to another only if we can show that the position rests on a more coherent set of specifications and interpretations of the common morality.[110]

In effect, common morality remains something "out there" regardless of the two differing positions on blood transfusion. The focus then turns toward whomever can articulate the best argument to demonstrate which party has the best claim to accord with common morality. This "solution by a better argument" procedurally or practically may seem productive in a case such as the Jehovah's Witness, but it must be noted that the impetus for judging the best argument stems from the current social majority. If Jehovah's Witnesses were the majority population, the common moral foundation B&C promote would surely need to shift to support withholding blood transfusions.

Likewise, when B&C anticipate examples critical of common morality, such as the "Pirates Creed of Ethics or Custom of the Brothers of the Coast" (1640), they can reject this on the grounds that such is a "particular morality" that does not accord with what is universally normative. The Pirate's Creed meets the need of "bare coherence" in that it is internally consistent, and yet for B&C the creed fails the test of universality. The rhetoric of pirates may seem ideal given that few if any consider it desirable to be robbed, pillaged, or murdered. The reality, however, as Arras reminds us, is that Rawls's process of reflective equilibrium is "maximally inclusive" such that all particular possibilities should be considered as the *commonness* of morality is being realized:

> If you do not like the way the process of reflective equilibrium is going, if you think that it currently overlooks some crucial pieces of the moral picture—such as a different moral outlook or a background theory of social stability—then this method simply asks you to toss it into the mix alongside all our other beliefs. Although the initial moral data of reflective equilibrium, i.e., our considered moral judgments, could conceivably be overturned—think, for example, of recent attitudes towards homosexuality—it is hard to imagine that most of them would or could be overturned in our lifetime; and if they are overturned, then it would no doubt be for the sort of good reasons

110. Beauchamp and Childress, *Principles of Biomedical Ethics* (8th ed.), 25.

that would lead Beauchamp and Childress to expand the scope of the norm of equal treatment within the common morality.[111]

In short, B&C seem optimistically to envision common morality as an ethical space in which one may navigate progressively ever more toward a realization of human moral flourishing, whatever that may be. However, "whatever that may be" is no small tag-along. As Lee points out, B&C offer an ethical approach that is "thick in status and thin in content" as it offers suggested practical solutions that are easily, even presumptively, confused as metaphysical truths. In the case of the Jehovah's Witness scenario, we are offered a prescriptive common moral solution, and yet the background for this prescription largely remains vacant. "To validate their theoretical position of recognizing the legitimate moral diversity in consistent with the prescriptive common morality, they still need to come up with another backup, if there is one."[112] Hence Lee rightly critiques B&C for offering a "thick" solution that is aimed at practical answers and yet "thin" in that it may support a multitude of different traditional and individual answers. Autonomy, nonmaleficence, beneficence, and justice, as B&C present them, are less suited to be ethical "answers" as much as the actual "grounds" for ethical debate. In the 2009 6th edition, B&C referred to the four principles first as the "thin set of four clusters of universal principles" but then expressed certainty that "near-complete agreement in moral judgment"[113] will arise from the common morality methodology. For the 2013 7th edition, however, the promise of principles was qualified yet firm: "Principlism, then, is not a mere list and analysis of four abstract principles. It is a theory about how principles link to and guide practice."[114] Given this confusion over the thick and thin status of principlism, Lee calls upon B&C either to assert (1) the Western or American "liberal individualist common moral intuition" that undergirds the application of principlism ("come take counsel with us!"), or (2) the real possibility that "no culture/tradition's usage of the four principles and conclusions based on them is better or worse" ("use our big four to your own taste!"). Can principlism be considered as anything else but "the method of common morality (moral objectivism) de jure but of moral relativism de facto?"[115]

In other words, B&C fail to identify their own rationale—or, perhaps more aptly, *principles*—for selecting universal principles of common

111. Arras, "Hedgehog and the Borg," 20.

112. Lee, "Problem of 'Thick in Status, Thin in Content,'" 527.

113. Beauchamp and Childress, *Principles of Biomedical Ethics (6th ed.)*, 25.

114. Beauchamp and Childress, *Principles of Biomedical Ethics (7th ed.)*, 25.

115. Lee, "Problem of 'Thick in Status, Thin in Content,'" 528.

morality. What is the measure for commonality amongst principles? Here it is clear that B&C wish to avoid entirely such a metaethical question and to remain in a quasifoundational middle sphere. Their emphasis on common morality as "a basic starter kit of abstract moral norms [is] meant to provide justificatory ballast to subsequent moral reflection."[116] B&C seek a pretheoretical, quasifoundationalism built upon presumably universal starting points, and consistent with this, they dedicate chapter 9 of *Principles* to surveying moral theories such as Kantianism, utilitarianism, rights theory, virtue ethics, etc., as a follow-up to their primary principles and as an explicit attempt to avoid inaugurating another "comprehensive ethical theory." Intending to limit their scope, B&C seek "to construct a coherent body of virtues, rights, principles, and rules for biomedical ethics."[117] In this way they bypass the essential foundational questions and instead focus on principles relevant for biomedical ethics "*drawn from* the territory of common morality, however small or large it may be." Their goal is not to survey the expansive boundaries of common morality itself; rather, the thesis of *Principles* "is merely that the principles and rules are a reasonable formulation of some vital norms of the common morality and that the principles we analyze are particularly suited to biomedical ethics."[118]

For some, this vague degree of foundational particularity may seem well and fine, but what remains in practice amidst this shifty language is a system of professionalized principles that can morph and bend at will, and the will behind this moral schema should not be ignored. This chapter has argued thus far that the morphing and shifting of principlist language will inevitably fall in line with whatever political power schema is currently *en vogue*. Rather than pursue explicitly a traditional or content-rich vision and *habitus* of the good, B&C enact a pragmatic process that adopts select terms and language such as autonomy, beneficence, nonmaleficence, and justice as thin threads dangling from past Enlightenment thinking. Discerning exactly what these terms entail and how they are employed is not easy. Articulating this point, Arras strikingly compares B&C to the character Borg in the science-fiction series *Star Trek: The Next Generation*.

> The Borg, a hive of cybernetically-enhanced humanoid drones, explore the universe in search of interesting new cultures and technologies, which they promptly conquer and incorporate

116. Arras, "Hedgehog and the Borg," 29.

117. Beauchamp and Childress, *Principles of Biomedical Ethics* (8th ed.), 385, 385–418.

118. Beauchamp and Childress, *Principles of Biomedical Ethics* (8th ed.), 445 (emphasis original).

> into their neural network en route to their goal of ultimate perfection. On encountering an alien culture, the Borg ominously announce, "Resistance is futile, you will be assimilated." Many of Beauchamp and Childress's critics know the feeling. No sooner do they launch a seemingly crippling broadside against the juggernaut of *PBE* from a casuist, narrativist, feminist, or pragmatist perspective than their critique is promptly welcomed with open arms, trimmed of its perceived excesses, and incorporated into the ever-expanding synthesis of the next edition.[119]

Beyond such a descriptive analysis, the fact remains that Beauchamp and Childress are *drawing* from "the territory of common morality" from within a secular vision of reality. In this way, B&C quintessentially represent the tendency in bioethics Engelhardt decried, the tendency to disregard the "difficulties that lie at the very roots of modern thought . . . [namely, proceeding] with the task of applying ethics as if it were obvious which secular ethic ought to be applied." Such a secular vision ignores the inevitable pluralism of postmodernity and the reality that "there is more than one vision of the cosmopolis and of the cosmopolitan." B&C are cosmopolitan liberals who must believe that they themselves possess the secular touchstone for what is "canonical, content-full, secular morality (and bioethics)," and hence the whole of their endeavor is "justifiable" independent of any "particular moral history and tradition." [120]

Considering B&C's *Principles of Biomedical Ethics* as something of a representation of bioethics itself, what emerges is a professionalized, easily accessed language capable of employment within various shades of liberal secularism. We are cast upon shifting tides, and where B&C may not intend to speak with "bioethical expertise" when "expertise" entails a biopolitical claim to power, the effect of their shifty language remains as a vacuum that inevitably will be filled. This vacuum, moreover, is a presence Milbank identifies within the "formal openness of liberalism which is designed to mitigate conflict, and an arbitrariness of content on the other hand—a 'positivism' which always threatens to overwhelm even the peace of mere suspended hostility which is the best that the *civitas terrena* can ever manage." The options for content within liberalism can be

> *either* "scientific" as in the case of eugenicism and the extermination of the supposedly weak (which happens in far more modes than we usually acknowledge) *or* it can be "religious" as in the case of recently emergent "fundamentalisms" which

119. Arras, "Hedgehog and the Borg," 12.

120. Engelhardt, *Foundations of Bioethics*, 9n16, 26–27.

> usually trade off, and theologically confirm, socio-economic liberalism, while also in certain strategic ways surpassing and opposing it.[121]

Milbank's broader thesis on liberalism is fitting for this biopolitical understanding of bioethics as he contends that the secular is "complicit" with an "ontology of violence."[122] The *civitas terrena* is fundamentally a matter of force and counter-force, and in the absence of any transcendent, ontological vision and *habitus* of the good, the expertise of a bioethicist is naturally rendered as a mechanized cog within some "arbitrary content." As Elliott warned, the "figure of the ethicist" may play a key role within Christopher Lasch's "culture of narcissism" bent on feelings or illusions of "well-being, health, and psychic security." This role will be "uniquely late-modern" in the turn away from priests or rabbis as the representatives of God. Those traditional "external figures" mean less today as the postmodern moral compass is directed "internally." Ironically, amidst this inward turn, bioethics draws us "to look outward again . . . to figures who derive their authority not from God but from a special kind of training. The authority of ethicists comes from their claim to expertise."[123]

In short, the discourses of bioethics reveal a pragmatic, principlist system of language that possesses its own self-referential rules, and those who master the language of bioethics gain the authority to be a "bioethicist" even while the foundation or grounding for this claim remains unclear. Bioethics is a tale of ethics lost in modernity, and we are left rudderless to navigate amidst the possible theories and principles of modern ethics. At this point, Wittgenstein's critique of philosophy grounded in theoretical or scientific thinking is strikingly appropriate, and the following chapter examines the life and philosophy of Wittgenstein to show how the prominent philosophical thinking employed in bioethics tends "to ask and answer questions in the way science does" (*BB*, 18). The confusion of such language evidenced in bioethics and modern, theoretical philosophy produces a confused and faulty metaphysics that the following chapter aims to uncover through Wittgenstein's critical and therapeutic practices with language.

121. Milbank, *Theology and Social Theory*, xi–xii (emphasis original).

122. Milbank, *Theology and Social Theory*, 4.

123. Elliott, "Tyranny of Expertise," 44.

2

Introducing Wittgenstein: Prophet of Modern Theorizing

The following is an exploration of Wittgenstein's metaphilosophy, a metaphilosophy that challenges the modern viewpoint that philosophy will provide "fundamental insights into the human condition and the ultimate character of the universe, leading to vital conclusions about how we ought to arrange our lives."[1] Wittgenstein's metaphilosophy may also be called his critique of philosophy, a critique of philosophy as a theorizing endeavor that I contend is employed within prominent discourses of bioethics. This theme in Wittgenstein is complex, and, naturally, interpretive "camps" abound, following different readings of Wittgenstein. This chapter does not purport to engage all of the vast secondary literature on Wittgenstein but rather to home in on Wittgenstein's metaphilosophy, and this entails a focus not on secondary arguments regarding tractarian insights that may be drawn for discourses of analytical positivism. The goal, rather, is to understand, for example, what Wittgenstein means by stating that the most important (and unwritten) point of the *Tractatus* is ethical and how the *Tractatus* may be seen as an "ascetical exercise in learning to acknowledge what may be said in order to respect that which is unsayable."[2]

1. Horwich, *Wittgenstein's Metaphilosophy*, 2.

2. Writing to Ludwig von Ficker regarding publishing the *Tractatus*, Wittgenstein stated: "I am pinning my hopes on you. And it will probably be helpful for you if I write a few words about my book: For you won't—I really believe—get too much out of reading it. Because you won't understand it; the content will seem really strange to you. In reality, it isn't strange to you, for the point of the book is ethical. I once wanted

Further, the goal here is to present an exposition of Wittgenstein's metaphilosophy that is consistent throughout his early and later works and yet reaches a fuller maturity in the *Philosophical Investigations*. The aim is not to critique or defend every part of his view but rather to describe it and then to demonstrate how Wittgenstein's critique of philosophy informs the discourses of moral epistemology in bioethics. In a real sense, the dilemmas with modern philosophical theorizing that Wittgenstein addresses are very much present within the moral calculus of bioethics. The first step in articulating Wittgenstein's metaphilosophy is to describe the cohesive, though nuanced, unity of early and late Wittgenstein. Bridging the "Wittgensteins" is a needed step for at least two reasons: (1) key to Wittgenstein's metaphilosophy and work on the whole, one should clarify how the structure and claims of the (early) *Tractatus* do or do not accord with the (later) *Investigations*; (2) and more significantly, I argue that the turn from the *Tractatus*'s emphasis on crystalline logic to therapeutic language in the *Investigations*[3] may be informative for bioethics.

1. Bridging the Wittgensteins

The work of the philosopher consists in marshalling recollections for a particular purpose (*PI*, 127).

Reading Wittgenstein can be arduous, not unlike navigating an alpine path suspended in fog. From the concise, analytical precision of the *Tractatus* to the cryptic aphorisms of the *Investigations*, Wittgenstein traverses puzzles in modern philosophy, and he himself is hardly immune from being puzzled, as evidenced by his lecture notes and letters to friends that reveal the dynamic process and formation of his thought. Amidst Wittgenstein's questions and puzzles over logic and language, however, we unquestionably glimpse a luminous mind at work. Bertrand Russell accepted Wittgenstein as a student poorly read in philosophy in October of 1911 only to announce

to give a few words in the foreword which now are actually not in it, which, however, I'll write to you now because they might be a key for you: I wanted to write that my work consists of two parts: of the one which is here, and of everything which I have not written. And precisely this second part is the important one. For the Ethical is delimited from within, as it were, by my book; and I'm convinced that, strictly speaking, it can ONLY be delimited in this way. In brief, I think: All of that which many are babbling today, I have defined in my book by remaining silent about it. Therefore the book will, unless I'm quite wrong, have much to say which you want to say yourself, but perhaps you won't notice that it is said in it" (Wittgenstein, "Letters to Ludwig Von Ficker," 94).

3. By "*Investigations*" here I mean Part I. I am following the practice of the revised 4th edition of Anscombe's translation of *Philosophische Untersuchungen* in describing Part I as *Philosophical Investigations* (1–693), Part II as *Philosophy of Psychology—A Fragment*.

in March of 1912 that he had nothing left to teach the prodigious Austrian. The protégé and master quickly swapped places, and such a keen intellect is surely one reason behind Bennett's comment in 1975: "Wittgenstein is long dead, and philosophers have recovered their nerve."[4]

The challenge of reading Wittgenstein is something he himself acknowledged. "[The *Tractatus*] is written for such men as are in sympathy with its spirit. This spirit is different from the one which informs the vast stream of European and American civilization in which all of us stand" (*PR*, 7). He considered his work and culture "foreign" to the predominant scientific and progressive modes of his contemporary culture (*CV*, 7), and he looked to figures as varying as Beethoven, Goethe, and Nietzsche who confronted "problems of the intellectual world of the West" in a prophetic way akin to his own (*CV*, 9).[5] In a similar vein, the preface of the *Tractatus* opens with Wittgenstein's remark that only those "who have themselves already thought the thoughts" expressed in the *Tractatus* will "perhaps understand it" (*T*, 27). What is this "spirit" readers must share to grasp Wittgenstein?[6] Doubtless, such statements shed light on the questioning reception of Wittgenstein in many philosophical circles committed primarily to analytical and discursive rationality. Or, likely even more troubling to many philosophers is the notion that a right end of philosophy is mystical silence: "There is indeed the inexpressible. This shows itself; it is the mystical (T 6.522). Whereof one cannot speak, thereof one must be silent" (*T*, 7). One must wonder also if the difficulty in reading Wittgenstein should be connected to Wittgenstein's difficulties in finishing his writing projects, especially later in his career. These respective difficulties, however, seem far removed from "prosaic obscurity" or from "writers block" and more due to the nature of philosophy itself and the ineffable to which Wittgenstein's philosophy points. Wittgenstein's understanding of philosophy as an activity, not as a series of philosophical propositions or theories, fundamentally makes it difficult to write a book articulating his method.[7]

4. Quoted in Edwards, *Ethics without Philosophy*, 2.

5. "If I say that my book is meant only for a small circle of people (if it can be called a circle), I do not mean that I believe this circle to be the elite of mankind; but it does comprise those to whom I turn (not because they are better or worse than others but) because they form my cultural milieu, my fellow citizens as it were, in contrast to the rest who are *foreign* to me" (*CV*, 10, emphasis original).

6. Frege puzzled over Wittgenstein's statement that the *Tractatus* would only be understood by those who have already "thought the thoughts which are expressed in it." Frege commented: "The pleasure of reading your book can therefore no longer be aroused by the content which is already known, but only by the peculiar form given to it by its author. The book thereby becomes an artistic rather than a scientific achievement; what is said in it takes second place to the way in which it is said" (as quoted in Monk, *Ludwig Wittgenstein*, 174).

7. Monk, *How to Read Wittgenstein*, 63.

The polarization of "younger" and "later" Wittgensteins is almost certainly the standard response to the puzzle of reading Wittgenstein, and often those who heavily lean on this distinction seem to cling primarily to but one of these poles, dismissing the other pejoratively as (early) "positivism" or (later) "mysticism."[8] Bertrand Russell exemplifies this latter tendency as he remarked of the later Wittgenstein: "[he] seems to have grown tired of serious thinking and invented a doctrine which would make such an activity unnecessary."[9] Whether or not Wittgenstein ever genuinely followed in the steps of Russell, seeking to emulate his method and positivist ambitions, is a strong matter of debate, yet it is clear that many see the Wittgensteins of the *Tractatus* and of the *Investigations* as seeking quite different goals.[10] Or, as Moyal-Sharrock proposes, a third Wittgenstein may be discerned primarily via the post-*Investigations* notes published now as *On Certainty*.[11] Such distinctions may be helpful amidst some types of Wittgensteinian scholarship, and yet for now it is more pertinent to look to the overarching unity of Wittgenstein. Admittedly, such a unity is likely for some the very point of contention in arguing for one or the other of the Wittgensteins, and yet I argue that the distinctions and reforms in Wittgenstein's thought that lead to the "early" and "later" labels exist within cohesive, overarching concerns, including (1) understanding the limits of philosophy, and (2) seeking a moral

8. John Verdi's wit renders this flat summary: "Ludwig Wittgenstein (1889–1951) was two of the most influential philosophers of the 20th century" (Verdi, "Doing Things with Words," 10).

9. From Russell's *My Philosophical Development*, quoted in Horwich, *Wittgenstein's Metaphilosophy*, 3.

10. Engelmann recounts Wittgenstein's "depression" after reading Russell's introduction to the *Tractatus*, the inclusion of which Engelmann thought was essential to getting the *Tractatus* published. That Wittgenstein was misread by one of his respected mentors likely contributed to his decision to take up elementary teaching, foregoing philosophy as a profession (See Engelmann, *Letters from Ludwig Wittgenstein*, 116–17). *The New York Review of Books* printed an interesting encounter between Allan Janik and D. F. Pears (Janik, "Wittgenstein's Strategy") on the question of Wittgenstein's alignment with Russell. Janik calls on Engelmann's *Letters from Ludgwig Wittgenstein* to question Pears's "mythical Wittgenstein, which was conceived by Russell, nurtured by the Vienna Circle, and sustained by a segment of Anglo-Saxon philosophers for the last forty-five-odd years" (para. 1). Janik and Engelmann see Wittgenstein in the tradition of language-mystics, utilizing tools of physics and logic to pursue "the same task as the language-mystics who maintained that language was incapable of expressing any sort of meaningful discourse concerning values" (para. 4). Furthermore, Monk identifies "the most important cause of [Wittgenstein's] depression as his failure to find a publisher for the *Tractatus*—or even a single person who understood it" (Monk, *Ludwig Wittgenstein*, 173).

11. Moyal-Sharrock, *Third Wittgenstein*.

vision of the "healthy human life."[12] Within this holistic view, Wittgenstein's admission in the preface to the *Investigations* of "grave mistakes" (*PI*, 4) in the *Tractatus* represents Wittgenstein's move from a representational theory of language that emphasizes logical necessity to a view of language defined by its activities and uses, but the change of emphasis from logic to language, however deeply significant, does not alter his overall goal of contesting abstract theorizing. The metaphilosophical concerns present in Wittgenstein's tractarian thinking change in form but not *in toto* as he moves to the *Investigations.*[13] Moreover, this reading accords with Wittgenstein's desire to publish the nearly completed *Philosophical Investigations* side-by-side with the *Tractatus Logico-Philosophicus* as he considered that the *Investigations* would only make sense in light of the *Tractatus.* Certainly, this was due in part to the corrections he wished to make but equally to the "continuities between the two." After all, Wittgenstein never denied or "repudiated" the philosophy of the *Tractatus.* He seemed, rather, to grasp more what was required in order "to do justice" to the philosophy of the *Tractatus*: namely, "that there can be no such thing as a philosophical proposition, and that philosophy was *entirely different* to science."[14]

If we do not restrict Wittgenstein's thought to static modes in two or three distinct periods, each with its respective masterpiece, we are more in agreement with Paul Engelmann, Wittgenstein's confidant and publisher, who points to the question of philosophy itself as key:

> We do not understand Wittgenstein unless we realize that it was philosophy that mattered to him and not logic, which merely happened to be the only suitable tool for elaborating his world picture. This the *Tractatus* accomplishes in sovereign fashion, ending up with implacable consistency by nullifying

12. Edwards, *Ethics without Philosophy*, 4.

13. More will be said below regarding Wittgenstein's self-criticism. For now, it is helpful to note that Horwich identifies eight theses from the *Tractatus* that are rejected in the *Investigations* (Horwich, *Wittgenstein's Metaphilosophy*, 96–104), and all are variations on Wittgenstein's transition from logic as the mirror-image or "reflection" of the world (*T*, 6.13, *kiene Lehre, sondern ein Spiegelbild der Welt*) with language simply representing this mirroring to his later acceptance of the complexity and interrelatedness of language as the more accurate way to see the world. Despite the critical transition from the earlier to later Wittgenstein, a core relatedness unites Wittgenstein's fascination first with crystalline logic and then with language that replaces logic. How so? Logic, language, and the world stand "in line," each one to another: "Other illusions come from various quarters to join the particular one spoken of here. Thought, language, now appear to us as the unique correlate, picture of the world. These concepts: proposition, language, thought, world, stand in line one behind the other, each equivalent to each" (*PI*, 96–97).

14. Monk, *How to Read Wittgenstein*, 64 (emphasis original).

> the result, so that the communication of its basic thoughts, or rather of its basic tendency—which, according to its own findings, cannot on principle be effected by direct methods—is yet achieved indirectly. He nullifies his own world picture, together with the "house of cards" of philosophy (which at that time at least he thought he had made collapse), so as to show "*how little is achieved when these problems are solved.*" What he wants to demonstrate is that such endeavors of human thought to 'utter the unutterable' are a hopeless attempt to satisfy man's eternal metaphysical urge.[15]

Engelmann's summary of Wittgenstein's consistent lifelong goal is striking: *how little is achieved when these problems are solved* and when theoretical philosophy has reached its end. Engelmann is not alone in reading a unified Wittgenstein. Horwich grants critical distinctions between the early and later Wittgenstein, while advocating an overall coherence through Wittgenstein's metaphilosophy, which may be defined as the investigation into the aims, methods, and boundaries of philosophy itself. Pointing to *PI*, 109—"philosophy is a battle against the bewitchment of our understanding by the resources of our language"—Horwich argues that key to Wittgenstein's message is the assertion that "a priori theories and sophisticated arguments" in support of "realms of phenomena" belonging to philosophers simply do not exist as most philosophers assume. "Philosophy cannot deliver the sort of knowledge that is usually regarded as its *raison d'être*," nor can philosophy "provide fundamental insights into the human condition and the ultimate character of the universe, leading to vital conclusions about how we ought to arrange our lives."[16] Given the modern expectations of what philosophy can and cannot do, Horwich reads Wittgenstein as something of a pessimist who receives a "curl of the lip" in many philosophical circles because of his claims that theoretical metaphysics is not substantive in itself but rather exists primarily as a perpetual conflict between the "required ideal" and the actually real (*PI*, 101–6).

> The more we examine actual language, the greater becomes the conflict between it and our requirement. (For the crystalline purity of logic was, of course, not something I had *discovered*:

15. Engelmann, *Letters from Ludwig Wittgenstein*, 96 (emphasis original). See *PI*, 118: "Where does this investigation get its importance from, given that it seems only to destroy everything interesting: that is, all that is great and important? (As it were, all the buildings, leaving behind only bits of stone and rubble.) But what we are destroying are only houses of cards, and we are clearing up the ground of language on which they stood."

16. Horwich, *Wittgenstein's Metaphilosophy*, 1–2.

> it was a requirement.) The conflict becomes intolerable; the requirement is not in danger of becoming vacuous.—We have got on to slippery ice where there is no friction, and so, in a certain sense, the conditions are ideal; but also, just because of that, we are unable to walk. We want to walk: but we need *friction*. Back to the rough ground! (*PI*, 107; emphasis original)

In the end, for Horwich, Wittgenstein is a "demystifying" philosopher concerned with "virtues of clarity and truth" as language problems are deflated, and the expectations for theoretical philosophy are lifted, chief amongst them being the "expectation that philosophy can and should deliver a priori theories that—like good scientific theories—are simple, unifying, and explanatory."[17] This summary of Wittgenstein's project is drawn primarily from the *Investigations* that in Horwich's eyes stands as the culmination of a "series of improvements" from the *Tractatus* to his writings in the thirties and beyond. "Therefore, we might well say that he produced one philosophy preceded by a variety of drafts."[18]

For Hacker, a unified Wittgensteinian conception of philosophy may be seen by turning to the radical noncognitivism of the *Tractatus* and *Investigations*, and he points out that there is "no philosophical knowledge" in either work as philosophy should be an "activity of elucidation by analysis." The *Tractatus* accepted ordinary language as standing in good logical order, allowing Wittgenstein to pursue facts and atomic facts as conceptions of the world within the effable necessity of logic. Even granting the crystalline purity of effable logic, however, Wittgenstein's overarching tractarian goal was not Frege's or Russell's positivism. Wittgenstein's move beyond categorical or formal concepts—i.e., object, property, relation, fact, proposition, color, number, etc.—leads to logical forms that stand before metaphysical pronouncements, such as, one is a number, red is a color, etc. The point is made explicit in *PI*, 114 & 116:

> *Tractatus Logico-Philosophicus* (4:5, emphasis original): "The general form of propositions is: This is how things are."—That is the kind of proposition that one repeats to oneself countless times. One thinks that one is tracing nature over and over

17. Horwich, *Wittgenstein's Metaphilosophy*, 211, 15. Here it might be helpful to point out that Horwich focuses primarily on Wittgenstein's critique of philosophical metaphysics at the expense of Wittgenstein's significant—albeit undeveloped—points on aesthetic and ethical reasoning (Edwards, *Ethics without Philosophy*, 75–103). Wittgenstein's aesthetic reasoning will be discussed below as an apt transition from his critique of theoretical philosophy to his support of apophatic and integral knowledge that assume a hierarchy in modes of knowledge.

18. Horwich, *Wittgenstein's Metaphilosophy*, xi–xii.

again, and one is merely tracing round the frame through which we look at it.

> When philosophers use a word—"knowledge," "being," "object," "I," "proposition/sentence," "name"—and try to grasp the *essence* of the thing, one must always ask oneself: is the word ever actually used in this way in the language in which it is at home?—What *we* do is to bring words back from their metaphysical to their everyday use.

"Tracing round the frame" in *PI*, 114 is Wittgenstein's way of reminding us that the pronouncements of categorical or formal concepts already assume a necessary logical form that enables such pronouncements in the first place. Hence Hacker points out that what such "pseudopropositions" aim to state is actually "shown by genuine propositions that contain number words, color names, or other names of objects." He adds:

> It is shown by features of the expressions in such propositions—namely, by the forms of the expressions—their essential combinatorial possibilities. These are represented by the variable of which the meaningful names are substitution-instances. An immediate consequence of this is that most of the propositions of the *Tractatus* that delineate the necessary forms of language and reality are nonsense. Hence Wittgenstein's penultimate remark in the book: "My propositions serve as elucidations in the following way: anyone who understands me eventually recognizes them as nonsensical, when he has used them—as steps—to climb up beyond them."[19]

In short, a core ambition in the earlier and later Wittgensteins is a non-cognitive conception of philosophy, a conception that for Hacker identifies the work of philosophy less with human knowledge and more with human understanding. Philosophy cannot be an empirical science, nor an a priori science. Put simply, "philosophy is no science" tasked with the challenge of new discoveries, let alone "arcane" truths, proofs, or "recherché 'entities' like universals." Philosophy is "a quest for understanding, not for knowledge."[20] Regarding his work on the whole, Wittgenstein wrote:

> It is all one to me whether or not the typical western scientist understands or appreciates my work, since he will not in any case understand the spirit in which I write. Our civilization is characterized by the word "progress." Progress is its form rather

19. Hacker, *Wittgenstein: Connections and Controversies*, 12.

20. Hacker, *Wittgenstein: Comparisons and Context*, 9–10.

> than making progress being one of its features. Typically it constructs. It is occupied with building an ever more complicated structure. And even clarity is sought only as a means to this end, not as an end in itself. *For me on the contrary clarity, perspicuity are valuable in themselves.* I am not interested in constructing a building, so much as in having a perspicuous view of the foundations of possible buildings. So I am not aiming at the same target as the scientists and my *way of thinking* is different from theirs. (*CV*, 7, emphasis added)

Drawing together the above assessments, what emerges in Wittgenstein is a metaphilosophy that contests theoretical and scientific philosophy, and here it is worthwhile to illustrate some of the context of such scientific philosophy. Hacker points to two holders of the Wykeham Professor of Logic Chair at Oxford, Michael Dummet and Timothy Williamson. Dummet remarkably claimed in 1978 that "philosophy has only just very recently struggled out of its early stage into maturity: the turning point was the work of Frege."[21] Likewise, in 2005, Williamson claimed that we have just now come to the "end of the beginning" of philosophy.[22] Such remarks seem to claim a lineage with modern pursuits of modes and methods of analysis to uncover "indubitable" foundations of knowledge. Descartes, Locke, Hume, and Kant, each in his own way, can be seen pursuing such human knowledge, which for Kant entailed placing metaphysics "upon the true path of science."[23] This path of science, according to Lacoste, is inevitably a reduction of something, such as the fragmenting of ancient *theoria* and *ethos* solely to "theoretical work," a path that leads to scientific visions of the world where

> *phusis* is handed over to the cares of physical science; logic alone remains. *Logos* alone remains, then, and it is no longer anything but the *logos* of logic and the surveillance that it exercises over scientific procedures. All philosophy faithful to its origins ought to protest against its reduction to the state of *Wissenschaft*: its proper rigor does not have to be that of the sciences, even if it ought not to be any less.[24]

The natural question, then, to ask of scientific philosophy seems clear: Where are the results of this philosophical knowledge? Bertrand Russell surveyed the state of philosophy and opined rather ambitiously that

21. Dummet, *Truth and Other Enigmas*, 457.
22. Williamson, "Must Do Better," 187.
23. Hacker, *Wittgenstein: Comparisons and Context*, 8.
24. Lacoste, *From Theology to Theological Thinking*, 11–12.

"philosophy, from the earliest times, has made greater claims, and achieved fewer results, than any other branch of learning [such that] the time has now arrived when this unsatisfactory state of affairs can be brought to an end."[25] Is philosophy a cognitive, theoretical discipline? Can philosophical questions, pursued via discursive a priori theories and analytical arguments, decipher knowledge from "realms of phenomena" whose study belongs specifically to the philosopher?

These are the sorts of notions and directions of thought—thinking we can easily align with B&C's principlism and pseudotranscendental common morality—that Wittgenstein seeks to challenge in both his early and later works. His concern is the "tendency," "attitude," and "craving" in philosophy for "generality" or commonality within entities that, in other words, manifests, for Wittgenstein, a wrongfully "contemptuous attitude toward the particular case."

> Our craving for generality has another main source: our preoccupation with the method of science. I mean the method of reducing the explanation of natural phenomena to the smallest possible number of primitive natural laws; and, in mathematics, of unifying the treatment of different topics by using a generalization. Philosophers constantly see the method of science before their eyes, and are irresistibly tempted to ask and answer questions in the way science does. This tendency is the real source of metaphysics, and leads the philosopher into complete darkness. I want to say here that it can never be our job to reduce anything to anything, or to explain anything. Philosophy really is "purely descriptive." (Think of such questions as "Are there sense data?" And ask: What method is there of determining this? Introspection? (*BB*, 18)

Wittgenstein's Neo-Kantian Grammar?

> *Thinking is surrounded by a nimbus.—Its essence, logic, presents an order: namely, the a priori order of the world; that is the order of possibilities, which the world and thinking must have in common.* (*PI*, 97)

> *"Language (or thinking) is something unique"—this proves to be a superstition (not a mistake!), itself produced by grammatical illusions. And now the impressiveness retreats to these illusions, to the problems.* (*PI*, 110)

25. As quoted in Hacker, *Wittgenstein: Comparisons and Context*, 8.

An important question in reading Wittgenstein—particularly regarding the dilemmas of modernity and scientific theorizing—concerns the relationship between his critique of language and Kant's critique of reason. Much is at stake in this question given the significance of Kant's constructive method upon all philosophy, and while it is clear that Wittgenstein is not an explicit follower of Kant, the question remains to what degree Wittgenstein works within a Kantian space.[26] In general, the *Tractatus* asserts that language is meaningful only through the representation of simple objects, and only by way of facts are propositions able to function rightly. Seemingly, these are "*a priori* conditions of the possibility of logic and language—of thinking (reasoning) and representing. And is that not a transcendental argument?"[27]

Hacker identifies at least six areas of convergence between Kant and Wittgenstein.

1. *Metaphilosophy*: both philosophers reflected heavily on the nature of philosophy itself; both agreed that philosophy (Kant's "pure philosophy") is distinct from natural or mathematical sciences.
2. *Dialectic*: both philosophers engaged the Dialectic of Reason—the "logic of conceptual illusion"—agreeing to certain systematic patterns of philosophical error and the "unquestioned assumptions underlying philosophical controversies."
3. *The bounds of sense*: both Kant and Wittgenstein engaged the bounds of sense; Kant sought "deduction" of a priori concepts, seeking the "conditions and limits" of their use in "judgments and experience" respectively. Wittgenstein investigated the conditions for the meaningful use of language.

26. Hacker has engaged in detail the relationship between the thought of Kant and Wittgenstein, and I am indebted to him on this topic. Following Hacker's survey of Wittgenstein's published and unpublished writings, as well as personal interactions with others, we cannot discern much regarding Wittgenstein's familiarity and engagement with Kant's work. Wittgenstein mentions Kant once in the pre-*Tractatus* notebooks (Wittgenstein, *Notebooks*, 19.10.14) and once again in the Tractatus 6.36111 regarding Kant's problem of right and left hands. In the *Nachlass*, Hacker traces two references to Kant (MS, 107, 183), noting that "in one, [Wittgenstein] writes that the limits of language show themselves in that the only way to specify the fact that agrees with a true proposition is to repeat the proposition. Here [in the *Nachlass* Wittgenstein] notes, 'we're concerned with the Kantian solution to the problems of philosophy'" (Hacker, *Wittgenstein: Comparisons and Context*, 31). Further, in 1931 Wittgenstein specifically listed Schopenhauer as an influence, yet Kant was not mentioned (MS, 154, 16r). Hacker summarizes: "The catch from the trawl is meager. It does not suggest any significant Kantian influence on Wittgenstein nor even any evident Kantian inspiration" (Hacker, *Wittgenstein: Comparisons and Context*, 31–32).

27. Hacker, *Wittgenstein: Comparisons and Context*, 31.

4. *Rationalism and empiricism*: both philosophers "repudiated foundationalist epistemology of the Cartesian or Lockean kind—our knowledge of how things are in the world around us is not *inferred* from how things sensibly seem to us to be."
5. *The nature of necessity*: Kant maintained that the "necessary truths of logic" are "entirely without content," while Wittgenstein argued that the "tautologies of logic are 'senseless' (i.e. have 'zero sense')." More generally, "both located the roots of non-logical necessity in us—albeit for different reasons and in a very different sense."
6. *Rational theology*: Kant and Wittgenstein alike "repudiated" rational theology.[28]

Crystallizing much of the above six confluences are phrases from Wittgenstein such as: "Men have always thought that there must be a sphere of questions whose answers—a priori—are symmetrical and united into a closed regular structure" (*T*, 5.4541). From Kant: "How else can we account for our inextinguishable desire to find firm footing somewhere beyond the limits of experience? Reason has a presentiment of objects which possess a great interest for it."[29] Conor Cunningham in particular suggests that Wittgenstein parallels Kant even to the point that "'language game' may well be a metaphysical term." He goes on to say:

> The operational *a priori* nature of the concept of language game comes further to the fore when Wittgenstein refers to the "framework on which the working of our language is based (for example in giving descriptions)," or declares that "the common behaviour of mankind is the system of reference by means of which we interpret"; meanwhile radical human concurrence "is agreement not in opinion but in way of life' (*PI*, 240; 206; 241).[30]

Cunningham contends that such an "*a priori* centrality of language" follows from Wittgenstein's description of the workings of language that enable "signification, intelligibility, and does so by regulatively demarcating sense from nonsense in whatever guise." Key for Cunningham's reading is that grammar is "structural" and provides rules that constitute sense, and the benchmark of grammar is the grammar itself: "The very fact that Wittgenstein stipulates that grammar is also nonarbitrary illustrates the *a priori* formalism of the term: 'If we could justify a rule, it could be violated.' It is

28. Hacker, *Wittgenstein: Comparisons and Context*, 33–34 (emphasis original).

29. Kant, *Critique of Pure Reason*, 629–30. See Kerr, "Metaphysics and Magic," 241–43.

30. Cunningham, "Wittgenstein after Theology," 76.

the transcendental nature of rules that renders justification unintelligible."[31] In short, for Cunningham, rules and grammar, alongside forms of life and language games (*PI*, 656), all function for Wittgenstein in a mode quite similar to Kant's categories. In seeking to validate human knowledge, the a priori is relocated in "meaning rather than experience," and while Wittgenstein's "semantic relativisation" seems nonmetaphysical, Cunningham instead sees in the later Wittgenstein an "*ad hoc* transcendentalism" where (1) certain constitutive categories or concepts are posited, and yet (2) an "indeterminate actuality" questions the veracity of categories and instead resides only ever "in what is shown (*zeigen*), that which comes out."[32] Despite the challenge of alternate statements from Wittgenstein such as "no part of our experience is also a priori . . . there is no order of things a priori" (*T*, 5.634) and "when someone else wishes to say something metaphysical . . . demonstrate to him that he has given no meaning to certain signs in his propositions" (T 6.53), Cunningham's Wittgenstein employs a neo-Kantian grammar in support of an *ad hoc* transcendentalism.

Drawing back from Cunningham's neo-Kantian reading of Wittgenstein, we can see premodern theological commitments against *philosophical* metaphysics—"the positing of reality other than God"—as the primary framework for Cunningham's arguments such that, broadly speaking, the Kantian discourse is most problematic in that it is nontheological and immanentized, finding its only point of possible stability within finitude.[33] Such commitments against philosophical metaphysics further raise the stakes for the ways we understand Wittgenstein aligning (or not) with Kant. Does Wittgenstein advocate—intentionally or unintentionally—a "philosophical metaphysics?" What is the extent of Kerr's claim that Wittgenstein's critique of language (*T*, 4.00431) is a "radicalization of Kant's critique of reason?"[34] It will be important to return below to Cunningham's broader reading of Wittgenstein advocating "transcendental vitalism" and "philosophical metaphysics,"[35] where I will make the case that Wittgenstein actually supports in part Cunningham's main thesis by *contesting* transcendental and metaphysical theorizing. First, however, further groundwork is needed regarding the way Wittgenstein and Kant employ transcendental arguments.

31. Cunningham, "Wittgenstein after Theology," 76.

32. Cunningham, "Wittgenstein after Theology," 77 (emphasis original).

33. Cunningham, "Wittgenstein after Theology," 86.

34. Kerr, *Theology after Wittgenstein*, 37. See Kerr, "Metaphysics and Magic," 241–43.

35. Cunningham, "Wittgenstein after Theology," 85–86.

For all the above points of convergence and similarities, vital differences emerge between Kant and Wittgenstein's use of transcendental arguments. In one sense, much of the way one reads Wittgenstein will turn on discerning these nuances, hence rendering Kerr's statement on Wittgenstein's "*radicalization* of Kant" as either an assertion of Wittgenstein's remarkable Kantian consistency or of his remaking and actual critique of the Kantian horizon.

The first point to recall is Wittgenstein's emphasis upon philosophical methodology that serves as a critical backdrop for grasping Wittgenstein. "Philosophy is a struggle against the bewitchment of our understanding by the resources of our language" (*PI*, 109). Likewise, Wittgenstein stated plainly: "[w]hat I am opposed to is the concept of some ideal exactitude given to us *a priori*, as it were. At different times we have different ideals of exactitude; and none of them is supreme" (*CV*, 37e). Systematic philosophy seeks advances through theories or theses, and this is directly opposed by Wittgenstein's metaphilosophy. "If someone were to advance theses in philosophy, it would never be possible to debate them, because everyone would agree to them" (*PI*, 128). What concerned Wittgenstein were not "realms of phenomena whose study is the special business of a philosopher and about which he or she should formulate profound a priori theories and sophisticated arguments in support of them."[36] Such philosophical discoveries, or the "unearthing of something," signifies the confusion that Wittgenstein was motivated to deflate. In seeking to deflate confusion amidst philosophical theories, however, Wittgenstein navigates amidst such theories that may lead to misreading him as favoring the very metaphysical conundrums he seeks to remedy:

> What is metaphysical there is not the content of some belief but *the laying down of a requirement*, the requirement of logical analysis. We do make sense, our propositions do stand in logical relations to each other. And such-and-such is *required* for that to be so. The metaphysics there is not in something other than language and requiring that it be like this or like that; *that* sort of metaphysics the *Tractatus* uses only ironically: it uses apparently metaphysical sentences, but in a way which is disposed of by the sentences which frame the book, in the preface and the final remarks. The metaphysics of the *Tractatus*—metaphysics not ironic and not cancelled—is the requirements which are internal to the character of the language as language, in their being a general form of sentence, in all sentences having this form. The

36. Horwich, *Wittgenstein's Metaphilosophy*, 1.

> metaphysics of the *Tractatus* is a kind of metaphysics that does not involve what is unsayably the case outside language . . . [37]

Reading Wittgenstein in such a light, we are able to discern more accurately how Wittgenstein employs the term "a priori" amidst a "philosophical *Weltanschauung*" at odds with Kant. Whereas Kant's essential concern is to investigate the possibility both of synthetic a priori judgments (B 19) and metaphysics as a science (B 22), Wittgenstein contested the "theorists of the past cultural period [who sought] to find the a priori where it isn't" (MS 183, 81). As Hacker has argued, the propositions of mathematics and logic for Wittgenstein are a priori, meaning that "apparently synthetic a priori propositions" such as "nothing can be red and green all over" or "time-travel is impossible" are a priori. Such a priori propositions are nonanalytic, however, and are not "correctly characterized as *knowing the truth of a description* of how things necessarily are in nature" as proper "a priori *descriptions* of the scaffolding of the world." To the contrary, no such scaffolding exists—"neither original (traditional metaphysics), nor constructed and imposed (Kantian metaphysics of experience). Such (apparently synthetic a priori) propositions *constitute* the scaffolding from *which we describe the world*."[38] At stake are two different visions of knowledge: Kant's synthetic a priori knowledge of the world purports to describe necessarily how things are, whereas Wittgenstein's knowledge is "*knowledge of rules of representation*."

> Consequently, Kant's question: 'How is synthetic a priori knowledge of nature possible?' crumbles in Wittgenstein's hands. . . . For, in Wittgenstein's view, what appear to be necessities of nature, and what Kant argued to be a priori principles that the understanding imposes upon intuitions to *constitute* nature, are no more than shadows cast upon nature by the grammar of our language.[39]

Beyond these considerations of philosophical worldview and intent for Wittgenstein's thought, a deeper consideration of transcendental arguments is in order, especially to engage a reading such as Cunningham's where Wittgenstein's may present an ad hoc transcendentalism. Looking first to the *Tractatus*, if we understand transcendental arguments in the general, logical sense of a *modus tollens*, then most certainly the *Tractatus* is transcendental—as would be Plato's or Aristotle's metaphysics and all other philosophers who have argued "from the nature of language or thought to

37. Diamond, *Realistic Spirit*, 19 (emphasis original).

38. Hacker, *Wittgenstein: Comparisons and Context*, 35 (emphasis original).

39. Hacker, *Wittgenstein: Comparisons and Context*, 36 (emphasis original).

the nature of the world."[40] This common and generalized form of transcendental argument, however, does not accord with Kant's specific notion of "transcendental proof." Kant's deduction, after all, relies upon the assertion that "the mode in which the manifold of sensible representation (intuition) belongs to one consciousness precedes all knowledge of the object as the intellectual form of such knowledge, and in itself constitutes a formal a priori knowledge of all objects, so far as they are thought (categories)" (A 129–30). Hence for Kant, any transcendental proof necessarily requires the existence of transcendental idealism, whereas the *Tractatus* denies the possibility of Kantian "pure" and "necessary" concepts of understanding emerging "only in ourselves" (A 130). Hence Hacker summarizes:

> 1. [The *Tractatus*] is not an attempt to vindicate the possibility of synthetic a priori knowledge. In fact, it denies that there is such a thing, since it holds that all expressible necessity is logical necessity. Insofar as there is metaphysical necessity it is inexpressible in propositions with a sense, and so cannot be the content of propositional knowledge.
> 2. The *Tractatus* does not link independent concepts by reference to possible experience and the a priori conditions of its possibilities. (Rather, it links material concepts with the formal concepts that are in effect variables of which the meanings of the material concepts are values.)
> 3. The *Tractatus* does not attempt to prove that the world of appearances (outer experiences) is a condition of the possibility of inner experience. That objects exist and that there are states of affairs is a presupposition of *logic*, not of experience.[41]

Wittgenstein's explorations in logic do not equate in the *Tractatus* to systematic reliance upon transcendental ideals in the precise Kantian sense. The opening factual claims of the *Tractatus* alongside the closing reminders of the ineffable and "what can only be shown" render a vital context for understanding the form of Wittgenstein's arguments: in a Kantian way, logic and language do present a necessary structure or scaffolding, and yet contra Kant this structure is not something to be scaled for knowledge of the truth of *how things are* in nature but rather for identifying our *norms of description*, our propositions from which we describe the world. The world and all its facts are contingent for "there are no rules or necessities in the natural world."[42] Hence the *Tractatus* states:

40. Hacker, *Wittgenstein: Comparisons and Context*, 45.
41. Hacker, *Wittgenstein: Comparisons and Context*, 46 (emphasis original).
42. Garver, "Philosophy as Grammar," 159.

> A necessity for one thing to happen because another has happened does not exist. There is only logical necessity (*T*, 6.37).
>
> As there is only a logical necessity, so there is only a logical impossibility (*T*, 6.375).

As these quotes intimate, the world and all its facts are not necessary in their states nor is it impossible for them to exist in another way. "'Strange coincidence, that every man whose skull has been opened had a brain!'" (*OC*, 207). Likewise, "I know = I am familiar with it as a certainty" (*OC*, 272). This contingency opens the possibility for Wittgenstein's metaphilosophy to clarify and seek understanding verses articulating first principles as knowledge. Wittgenstein's concern is not foundationalist; his goal, rather, is to identify necessity and impossibility where it *is* present in the human sphere of logic and language. The "scaffolding of our thoughts" that forms our way of looking at things arises from our inherent experience in the human realm; Wittgenstein notes obviously that just as every human being has parents, so, too, thoughts grow from within the context of experienced life (*OC*, 211). Such conceptual scaffolding naturally bears a Kantian reminiscence, and yet such forms arise from communal experience and hence lack a true transcendental origin.

To better understand Wittgenstein's focus on necessity from within human experience and logic, we can look to his developing thought around the year 1929 when he was engaging Schlick, Waismann, and others within the Vienna Circle regarding how the *Tractatus* might or might not align with their positivism. According to Monk, Wittgenstein began during this time to attend to "a more comprehensive syntax" that functioned amidst the inner connections of logical propositions, and he saw this syntax exemplified in the geometry of a circle. Imagine someone posits that a "circle" has a length of 3 cm and a width of 2cm. The natural response of course is to ask what one means then by a circle? "These rules are provided by the syntax, or, as Wittgenstein also says, the 'grammar' of our language, which in this case establishes an 'internal connection' between something's being a circle and its having only one radius" (*VC*, 78).[43] Wittgenstein's point was that the rule for a circle having only one radius arises from the syntax or the "grammar" of our language and not some other presumed a priori necessity of "what is" a circle. By thus turning the question, Wittgenstein sees inference emerging from the syntax of systems of propositions agreed upon within the human realm just as the syntax of human agreement regarding colors distinguishes between, say, red and

43. Monk, *Ludwig Wittgenstein*, 285.

blue. Were someone to say a red ball is blue, the natural response would be to question this statement based on the agreed-upon definition of the two colors. Why make this distinction? What is, in effect, the real difference between the syntax of agreed-upon propositions and "*necessary* properties of space, time, and matter?" Wittgenstein's answer addresses contingency and demonstrates his phenomenological approach (distinguishing sense from nonsense) versus one of physics (determining true or false): "Physics does not yield a description of the structure of phenomenological states of affairs. In phenomenology it is always a matter of possibility, i.e. of sense, not of truth and falsity" (*VC*, 63, emphasis original). Wittgenstein's thought was very much in development at this time, and yet, contra the positivists, he landed firmly on the necessity of determining the syntax and sense of things. Monk draws attention to Wittgenstein's distancing himself from a positivistic "dogmatic application" of a theory of meaning:

> I used at one time to say that, in order to get clear how a sentence is used, it was a good idea to ask oneself the question: "How would one try to verify such an assertion?" But that's just one way among others of getting clear about the use of a word or sentence. For example, another question which it is often very useful to ask oneself is: "How is this word learned?" "How would one set about teaching a child this word?" But some people have turned this suggestion about asking for the verification into a dogma—as if I'd been advancing a theory about meaning.[44]

Implicit within Wittgenstein's aim to uncover the syntax and sense of things is a certain givenness in language, evidenced empirically, that differs notably from Kant's ideal language. Hence in Garver's reading, Wittgenstein can be seen presenting a generalization of Aristotle's categories, especially where Aristotle indicates varieties in truth-claims. Wittgenstein picks up significantly on puzzling variations in language—jokes, orders, exclamations, etc.—that constitute a phenomenon to be experienced and reflected upon.[45] This givenness of language is not that it *has* to exist but that it *does* as a "feature of human life" and a "very general fact of nature." Wittgenstein assumes no "ultimate or metaphysical value" in the rules of grammar, à la transcendental reasoning, but rather that these rules are social facts simply given to us.[46]

44. Monk, *Ludwig Wittgenstein*, 287–88; see also 283–85.

45. At one point (1929–?), Wittgenstein seemed to ponder phenomenology as a parallel to his methods with grammar. Spielgelberg, "Puzzle of Ludwig Wittgenstein's *Phänomenologie*," 244–56. See Garver, "Philosophy as Grammar," 146–47.

46. Garver, "Philosophy as Grammar," 156–57. See also *OC*, 211.

> Our mistake is to look for an explanation where we ought to regard the facts as "proto-phenomena." That is, where we ought to say: this is the language game that is being played. The point is not to explain a language-game by means of our experiences, but to take account of the language-game. What is the purpose of telling someone that previously I had such-and-such a wish?—Regard the language-game as the primary thing. (*PI*, 654–56)

The *Tractatus* has been the primary emphasis to this point, and yet Wittgenstein's transition from logic to grammar in the *Investigations* further clarifies the nature of Wittgenstein's Kantian radicalization. The turn to grammar in the *Investigations*—Wittgenstein's Copernican revolution—can be seen as a transition from "necessary, but ineffable, truths of metaphysics . . . presupposed by logic" to a "grammar of discourse concerning events." This transition can be visualized by recalling Wittgenstein's puzzles with color: to admit that red is darker than pink is to know a rule of inference, and that inference itself plays a role in how we conceive of "red" and "pink." To acknowledge that "red is darker than pink is not to know a synthetic a priori proposition about nature, but rather to know a rule."[47] The grammatical rule of naming colors functions via consent amidst a community and differs significantly from propositional knowledge about nature. In a similar way, the law of causality is not, for Wittgenstein, an absolute necessity existing autonomously in nature awaiting cognitive discovery; *causality* is the wrong question and approach if we accept Wittgenstein's critique of synthetic a priori propositions as norms of representation.

> If the law of causality is indeed part of our form of representation, then to know that every event has a cause is just to know that if something is described as being an event, if may be inferred that it had a cause. If the inference from "E is an event" to "E was caused" is a priori legitimate, that would be because it is partly constitutive of the meaning of "event." What Kant thought to be synthetic a priori truths describing necessary constraints upon reality are merely what Wittgenstein called "grammatical propositions" seen through a glass darkly. A grammatical proposition is a rule of representation in the guise of a description of how things necessarily are.[48]

Wittgenstein's emphasis on "how things necessarily are" is a key assumption and demonstration of his willingness to ignore forms of skepticism that seek to prove the existence of anything. Rooted in grammar,

47. Hacker, *Wittgenstein: Comparisons and Context*, 47.

48. Hacker, *Wittgenstein: Comparisons and Context*, 48.

Wittgenstein's philosophy is methodological, an "empirical project based on use." Whereas Kant's critique can be described as "an ideal language project based on analysis of propositions or judgments," Wittgenstein's empirical and descriptive method follows the phenomenon of language to uncover components of life within natural history. In this way, the critical theme of Wittgenstein's grammatical philosophy emerges again as he shuns speculative or dogmatic metaphysics in favor of a more "humble moral achievement" where philosophy may be pictured as a servant of the sciences seeking "conceptual clarity and argument" rather than an epistemic discipline undergirding them.[49]

Are there transcendental arguments in the *Investigations*? The form of generalized transcendental arguments are discernable in the *Tractatus*, and yet Wittgenstein's later explorations with language make clear that he employs a Kantian form even as a critique of Kant.[50] Wittgenstein's task with philosophy is not constructive in terms of theory or metaphysics but rather critical in the sense of clarifying confusions amidst how our questions are phrased. Kant asks: "How is synthetic a priori knowledge of nature possible?" The assumption behind this question is that prior to, or independently of, experience the natural world necessarily is conceivable. Hence if this question itself is unquestioned, Kant's correlation between knowledge of metaphysical judgments and the necessary conditions of this judgment seems "ingenious" and "profoundly compelling." Countering such a correlationist move, Wittgenstein turns to the question of what seems to be a natural necessity. "Essence is expressed in grammar. Consider: 'The only correlate in language to an objective necessity is an arbitrary rule. It is the only thing which one can milk out of this objective necessity into a proposition'" (*PI*, 371 & 372). Hacker comments:

> What appear to be necessary and universal truths about the world are norms of representation. They are not expressions of knowledge of necessities constitutive of the realm of nature, but rather rules for the use of words in the guise of descriptions. They are not rules for nature, but rules for the description of nature. They are grammatical propositions. Wittgenstein's account of the nature of such propositions is wholly independent of his account of the conditions of the possibility of self-ascription of experiential predicates (or, more accurately, of avowals of

49. Garver, "Philosophy as Grammar," 167.

50. Wittgenstein may critique Kant most directly on the concepts of apperception, consciousness, and self-consciousness. See "Kant's Transcendental Deduction," in Hacker, *Wittgenstein: Comparisons and Context*, 54–77.

> experience). This is of capital importance. So, Kant and Wittgenstein take different paths through the conceptual jungle.[51]

The answer to the question of how Wittgenstein relates to Kant is nuanced, requiring careful qualifications for possibly describing Wittgenstein's arguments as transcendental. The reading of Wittgenstein given here sees his radicalization of Kant as a qualified employment of Kant's method that importantly rejects transcendental arguments aimed at constructing philosophical theories. Doubtless, Wittgenstein aligns logic, language, and the world as though "equivalent," and yet his project is far from seeking metaphysics as a science of being (*PI*, 96–97).[52] What is more important to Wittgenstein than transcendental arguments is the possibility of the transcendent, which lies beyond philosophy as a discipline of limiting reason and deflating superstitions (*PI*, 110). Writing to Engelmann in 1919, Wittgenstein stated, regarding the *Tractatus*:

> The book's point is an ethical one. I once meant to include in the preface a sentence which is not in fact there now but which I will write out for you here because it will perhaps be a key to the work for you. What I meant to write, then, was this: My work consists of two parts: the one presented here plus all that I have *not* written. And it is precisely this second part that is the important one. My book draws limits to the sphere of the ethical from the inside as it were, and I am convinced that this is the ONLY *rigorous* way of drawing those limits. In short, I believe that where *many* others today are just *gassing*, I have managed in my book to put everything firmly into place by being silent about it. And for that reason, unless I am very much mistaken, the book will say a great deal that you yourself want to say. Only perhaps you won't see that it is said in the book. For now, I would recommend you to read the *preface* and the

51. Hacker, *Wittgenstein: Comparisons and Context*, 64.

52. Kerr, "Metaphysics and Magic," 242. The first part of *PI*, 97 intimates a correlation between logic, language, and the world, and yet the latter half of *PI*, 97 explains the nature of this correlation:

> We are under the illusion that what is particular, profound, and essential to us in our investigation resides in it trying to grasp the incomparable essence of language. That is, the order existing between the concepts of proposition, word, inference, truth, experience, and so forth. This order is a *super*-order between—so to speak—*super*-concepts. Whereas, in fact, if the words "language," "experience," "world" have a use, it must be as humble a one as that of the words "table," "lamp," "door" (emphasis original).

conclusion, because they contain the most direct expression of the point of the book.[53]

These are significant words for reading Wittgenstein, positing a challenge for those who align Wittgenstein within the context of logicians such as Frege and Russell or the Vienna Circle positivists. Employing tools learned from his studies in physics and logic, Wittgenstein's philosophy is a discipline of critique for the sake of transcendent or higher values. In this sense, pace Wittgenstein's qualified Kantian paradigm, the more apt final context for his project lies with the tradition of language-mystics who see the shortcomings of language before the ineffability of the mystical (*T*, 6.522). Wittgenstein's *via negativa* is hence based on a posture of humility as he seeks to map the "boundaries of language," and, paradoxically, such a boundary is discovered in part through language: "the inclination, the running up against something indicates something. St. Augustine knew that already when he said: What, you swine, you want not to talk nonsense! Go ahead and talk nonsense, it does not matter!" (*VC*, 69).[54]

Moreover, it is precisely amidst this approach of speech engaging the boundaries of language that we uncover an important insight regarding Wittgenstein's view of philosophy in relation to theology. Wittgenstein's Galilean "kink" in the history of philosophy questions a type of theology that seeks metaphysical support from philosophy. Fergus Kerr maintains that such natural theology may be seen in theologians from Origen to Thomas Aquinas, and Kerr hence suggests that theologians from Luther to Barth who see philosophy as a threat may find Wittgenstein's philosophy

53. Engelmann, *Letters from Ludwig Wittgenstein*, 143–44 (emphasis original). The similarity between this letter to Engelmann and the letter to von Ficker quoted at the opening of this chapter further verifies Wittgenstein's emphasis on seeking the ethical transcendent rather than transcendental metaphysics.

54. During Wittgenstein's "verificationist phase" (ca. 1929), Monk describes Wittgenstein's interest in writing an autobiography and the role his words would play as a discipline or practice of enhancement: "Any autobiography he might have written would almost certainly have more in common with St. Augustine's *Confessions* than with, say, Bertrand Russell's *Autobiography*. The writing of it would, that is, have been fundamentally a spiritual act. He considered *Confessions* to be possibly 'the most serious book ever written'. He was particularly fond of quoting a passage from Book I, which reads: 'Yet woe betide those who are silent about you! For even those who are most gifted with speech cannot find words to describe you', but which Wittgenstein, in discussing it with Drury, preferred to render: 'And woe to those who say nothing concerning thee just because the chatterboxes talk a lot of nonsense' . . . [Such] free translations, even if they fail to capture Augustine's intended meaning, certainly capture Wittgenstein's view. One should put a stop to the nonsense of chatterboxes, but that does not mean that one should refuse to talk nonsense oneself. Everything, as always, depends on the spirit in which one does it" (Monk, *Ludwig Wittgenstein*, 282).

advantageous.[55] Kerr's point is that from 1930 Wittgenstein sees philosophy consumed with assumptions and problems of language. Kant's prime question—"How is synthetic a priori knowledge of nature possible?"—enacts numerous language errors and theoretical constructs that need to be disassembled. "The problems are solved, not by coming up with new discoveries, but by assembling what we have long been familiar with. Philosophy is a struggle against the bewitchment of our understanding by the resources of our language" (*PI*, 109). Likewise, "philosophy, as we use the word, is a fight against the fascination which forms of expression exert upon us" (*BB*, 27). Wittgenstein's rhetoric of "bewitchment" and "fascination" arises from philosophical conundrums with unquestioned questions such as "What is time?" or "What is knowledge?" or "What is a number?" Questions such as these are normally asked with an expectation that the answer will conform to "strict rules" as the philosophically puzzled man "sees a *law* in the way a word is used" and yet finds ambiguity and paradox in the question's answers (*BB*, 27, emphasis added). Hence the mistakes that Kant would attribute to reason, calling for philosophical critique, would for Wittgenstein be more rightly seen as superstitions and "grammatical illusions" (*PI*, 110) rooted in misused similes (*PI*, 112), propositions *PI*PI, 114), and/or everyday conversational language (*PI*, 117) that needs to be reclaimed "from their metaphysical to their everyday use" (*PI*, 116).[56]

Much more could be said, but in this introduction to Wittgenstein's thought, what most emerges is a prophetic vision of philosophy in the modern world, a modern world that puzzles over principles as *laws* that are born along via transcendental language. As we will see, this use of language is nothing less than a dilemma in modernity furthered by the spirit of scientific theorizing. In the following chapter, our next step is to engage Wittgenstein's critiques of scientific theorizing through (1) language games and (2) Wittgenstein's prioritization of an ethical transcendent versus transcendental pseudoethical language. The maneuvers of secular bioethics and transcendental language share a strong family resemblance in modernity as both seek "to keep something higher spellbound" in language (RF, 116–17), and bioethics stands to glean much from Wittgenstein's push toward ethics without (theoretical) philosophy.

55. Kerr, "Metaphysics and Magic," 250.

56. See *BB*, 25–28; Kerr, "Metaphysics and Magic," 250–54.

Part II

Critiquing Scientific Theorizing

3

Collapsing Theory: The Impulses of Scientific Thinking

Language as Therapy

> *The ideal, as we conceive of it, is unshakable. You can't step outside it. You must always turn back. There is no outside; outside you cannot breathe.—How come? The idea is like a pair of glasses on our nose through which we see whatever we look at. It never occurs to us to take them off.* (*PI*, 103)

To speak of therapy is to assume a disease or illness, and in the context of Wittgenstein, language as therapy applies to the sickness of misplaced expectations for what theoretical philosophy can offer. Language as therapy, then, is another way of describing Wittgenstein's critique of a particular type of theoretical philosophy (*PI*, 133). The goal now is to move deeper into Wittgenstein's overarching goal and metaphilosophy by articulating the therapeutic critique that stems from his view of language. Language in the *Investigations* does not guide us from one theory or science to another deemed more suitable; the aim is, rather, to remove the glasses of theory from our nose entirely. Wittgenstein seeks to turn our *Weltanschauung* (*PI*, 122) away from a fascination in "that order, the ideal"

(*PI*, 105) that floats above reality.[1] Actual language, for Wittgenstein, is the way to draw us to firm ground once again (*PI*, 107).

Drawing a summary sketch of Wittgenstein's critique via language is not easy, but, following the lead of Fogelin, it is helpful to trace at least two key themes—referentialism and logical perfectionism—that speak well for the whole.[2] Defined simply, referentialism is the presumptive view that words necessarily refer to things. Referentialism first and foremost is *presumptive* because it seems naïve on the one hand to think that *all* words stand for things or objects, and that *all* sentences represent the arrangement of things with one another, and yet this presumptive move marks Wittgenstein's choice "to ignore the obvious truth" by associating names with objects.[3] Undoubtedly, this critique is aimed at the philosophical confusion that Wittgenstein himself succumbs to in the *Tractatus*, and the opening of the *Philosophical Investigations* contains two classic examples of this, first from Augustine's *Confessions* and then with the shopkeeper's five apples, both of which make a point regarding the essence of language (*PI*, 1). Wittgenstein contemplated opening the *Investigations* with an allusion to Goethe—"what an intelligent man knows is hard to know"—via a description of nature (*CV*, 11) or with the difficult line "all is in flux" (*CV*, 8), and yet in the end he could not but open with a reference from his beloved *Confessions*.[4] This passage from *Confessions* I, 8 needs little introduction as Augustine traces the infant's process of connecting words to things as well as intentions to bodily expressions, sounds, and movements. Kerr takes the *Confessions*' opening as an "egocentric perspective" where the self is depicted as "inside," a sort of Cartesian posture Wittgenstein is critiquing. "Thus . . . Wittgenstein was placing his explorations of the epistemological predicament of the self in the context of a narrative which, as it interweaves biblical language with

1. For Wittgenstein, a fascination with ideals inevitably affects one's life. According to Monk, "this is especially clear when we consider that in the case of a philosophical *Weltanschauung* the consequences of a 'change of aspect' might be a change of *life*. In Wittgenstein's case, the consequence—'the outward criteria'—that he earnestly hoped for was a culture which treated music, poetry, art and religion with the same respect and seriousness with which our present society treats science" (Monk, *How to Read Wittgenstein*, 516, emphasis original).

2. Fogelin, "Wittgenstein's Critique of Philosophy," 34–58.

3. Fogelin, "Wittgenstein's Critique of Philosophy," 37. See also Monk, *How to Read Wittgenstein*, 73.

4. Kerr points to von Wright's comment regarding Wittgenstein's appropriation of Augustine's style in the *Confessions*: "The philosophical sections of St Augustine's *Confessions* show a striking resemblance to Wittgenstein's own way of doing philosophy" (Kerr, *Theology after Wittgenstein*, 39). Additionally, see Monk, *Ludwig Wittgenstein*, 282.

metaphysical dualism, autobiography with doxology, establishes the sense of the 'I' in the sight of God which remains the paradigm for the self even where the theology has been abandoned."[5]

Kerr persuasively traces the theological epistemology of the self specifically through Wittgenstein, and yet arguably the main question hinges on what point exactly the *Confessions* passage makes regarding referentialism. According to Monk, it is sometimes thought that the *Confessions* quote reveals Augustine's *theory* of language that Wittgenstein then challenges.[6] Rather than positing a theory, however, Augustine's autobiographical *Confessions* may be seen as revealing a dynamic process of learning language, a system where definitions and rules are learned through gestures and expressions. Wittgenstein certainly takes Augustine's picture of language as an introduction to referentialism—"the words in language name objects—sentences are combinations of such names" (*PI*, 1)—and yet the passage is not simply something Wittgenstein critiques. After all, Augustine does not refer solely to language in the abstract but rather "we might say, does describe a *system* of communication" (*PI*, 3, emphasis added). Augustine is describing the way that he learned to talk, and this paints for us a picture or metaphor (*PI*, 115). Further, Wittgenstein never states openly any disagreement with the *Confessions* passage, even while critiquing the issue of referentialism that the passage effectively brings to light.

What is clearer is the second example of the shopkeeper's five red apples, where we are reminded in an extremely simplified way that the *use* of words such as "red" and "five" surpass explanations. In *PI*, 2, Wittgenstein states explicitly that "use" of words is a "philosophical notion of meaning [that] is at home in a primitive idea of the way language functions." Furthering Augustine's picture of language, we should imagine builder A and an assistant B who are passing along stones that need further names:

> A is building with building stones: there are blocks, pillars, slabs and beams. B has to pass him the stones and to do so in the order in which A needs them. For this purpose they make use of a language consisting of the words "block," "pillar," "slab," "beam." A calls them out; B brings the stone which he learnt to bring at such-and-such a call. -Conceive of this as a complete primitive language. (*PI*, 2)

In short, two core claims or strategies of language as therapy are present in the opening paragraphs of the *Investigations* that are then developed and described through the rest of the work. Firstly, the referentialist picture

5. Kerr, *Theology after Wittgenstein*, 42.

6. Monk, *Ludwig Wittgenstein*, 364.

where individual words name objects—without noting "differences between types of words" (*PI*, 1)—leads to philosophical confusion as words are used uncritically. Secondly, Wittgenstein's pictures of prephilosophical, or primitive, language are intended as ways to help "dig out philosophical confusion by its prephilosophical roots." The emphasis here is on the phenomenon of *learning* language rather than purportedly intuiting language abstractly—as though one could eat the fruit of a tree that had no roots. Wittgenstein describes instruction in language unfolding through a process where "the learner names the objects; that is, he utters the word when the teacher points at the stone.—Indeed, there will be an even simpler exercise: the pupil repeats the words after the teacher—both of these being speech-like processes" (*PI*, 7). This educating (*ēdūcō*: to lead, conduct) process Wittgenstein calls "language-games," a term generally parallel to primitive language.

Some may object that Wittgenstein's primitive language oversimplifies the matter, and that the referentialist picture of language (*PI*, 2) and the builder-slab language game (*PI*, 8) examples "consist only of orders" and hence are selective examples. The point, however, seems to be that these examples indicate the roots of the problem and solution such that if we grasp these opening overtures, it will be possible to move beyond simple examples to the further complications and systems added into our language, even such as "the symbolism of chemistry and notation of the infinitesimal calculus." Wittgenstein compares language to "an ancient city: a maze of little streets and squares, of old and new houses, of houses with extensions from various periods, and all this surrounded by a multitude of new suburbs with straight and regular streets and uniform houses" (*PI*, 18).

Moreover, the dynamic configurations of language seem to help indicate two problems, particularly with the referentialist picture of language in regards to naming. Naming does not account for all uses of language, and referentialism does not rightly account for *naming* itself.[7] "Think just of exclamations, with their completely different functions [than referring to *things*]: Water! Away! Ow! Help! Splendid! No! Are you still inclined to call these words 'names of objects?'" (*PI*, 27). Likewise, in *PI*, 39 and 40, Wittgenstein ponders what happens when the simple object "behind" a name is fundamentally altered or lost. The sword named "Nothung" represents parts combined in a particular way, and yet when the sword is shattered and lost, the name Nothung maintains a "*sense*" of meaning. Likewise, when Mr. N. N. dies, it is only right to say that the "*bearer*" of the name has died and not the entire "*meaning*." In short, Wittgenstein's *Investigations* seeks to chip away at the notion that the philosophical analysis of language could reveal

7. Fogelin, "Wittgenstein's Critique of Philosophy," 39.

the underlying order within both logic and the world (*T*, 5.5563). Much of the *Tractatus* fell into this error, and yet even amidst such analytical philosophizing (*T*, 4.002), Wittgenstein was not seeking transcendental a priori truths of the world in the traditional sense, but rather even through and amidst such "grave mistakes" (*PI*, preface) he was seeking the "something that pervades everything sayable—something itself unsayable."[8] Hence Wittgenstein's later investigations clarify his metaphilosophy as a critique of metaphysics seeking "complete exactness" and purporting to penetrate theoretically into the essence of things (*PI*, 91):

> This finds expression in the question of the *essence* of language, of propositions, of thought.—For although we, in our investigations, are trying to understand the nature of language—its function, its structure—yet *this* is not what the question has in view. For it sees the essence of things not as something that already lies open to view, and that becomes *surveyable* through a process of ordering, but as something that lies *beneath* the surface. Something that lies within which we perceive when we see *right into* the thing, and which an analysis is supposed to unearth. "The essence is hidden from us": this is the form our problem now assumes. We ask: "*What* is language?," "*What* is a proposition?" And the answer to these questions is to be given once for all, and independently of any future experience. (*PI*, 92, emphasis original)

The analytical and theoretical questioning "what?" is an illusion. "Thought, language, now appear to us as the unique correlate, picture of the world," and the assumption is that "proposition, language, thought, [and] world" all stand in line as equivalent each to each. But something is missing: what are these words "to be used for now?" The language game as a "loose set of descriptions" is missing (*PI*, 96).[9]

Wittgenstein's critique of referentialism is perhaps clearest in terms of talk about imaginative or mental thoughts, especially when imagining someone else's pain (*PI*, 293–315). "Could someone who had never felt pain understand the word 'pain?'" (*PI*, 315). The difficulty lies in assuming mental processes can connect person A's physical pain with person B's. This may lead to statements such as "I can only *believe* that someone else is in pain, but I know it if I am" (*PI*, 303, emphasis original) because it is difficult "to imagine someone else's pain on the model of one's own." Wittgenstein continues:

8. Kerr, "Metaphysics and Magic," 242.

9. Fogelin, "Wittgenstein's Critique of Philosophy," 41.

> Yes: one can resolve to say "I believe he is in pain" instead of "He is in pain." But that's all.—What looks like an explanation here, or like a statement about a mental process, in truth just exchanges one way of talking for another which, while we are doing philosophy, seems to us more apt. Just try—in a real case—to doubt someone else's fear or pain! (*PI*, 303)

The mental transition from knowing to believing pain illustrates the puzzle. If we treat all language the same, assuming that it "always functions in one way," serving the same "purpose," be it language about houses, pains, good and evil, or whatever, then we tend to be forced into a paradox by grammar:

> "But you will surely admit that there is a difference between pain behaviour with pain and pain-behaviour without pain."—Admit it? What greater difference could there be?—"And yet you again and again reach the conclusion that the sensation itself is a Nothing."—Not at all. It's not a Something, but not a Nothing either! The conclusion was only that a Nothing would render the same service as a Something about which nothing could be said. We've only rejected the grammar which tends to force itself on us here. (*PI*, 304)

For Wittgenstein, to generalize from one case—my experience of pain—to the claim that I know what the word "pain" means is a great irresponsibility. "Well, everyone tells me that he knows what pain is only from his own case!" (*PI*, 293).[10] Wittgenstein compares this conundrum to everyone holding a box with "something in it which we call a beetle." No one can peer into the other boxes, and all claim to know what a beetle is by virtue of his box's contents despite the possibility that all have something different in their respective boxes. The point is clear that "the thing" in the box evades the grounding of a language game, perhaps most so if the box is empty, for then the thing would not even be a "Something." "That is to say, if we construe the grammar of the expression of sensation on the model of

10. If pain is not clearly a representational thing, Wittgenstein suggests that words may refer to sensations. Hence, stating one is "in pain" is an expression but not a description of pain. Hence, Wittgenstein ponders: "How do words refer to sensations? . . . How does a human being learn the meaning of names of sensations? For example: the word 'pain.' Here is one possibility: words are connected with the primitive, natural, expressions of sensations and used in their place. A child has hurt himself and he cries; then adults talk to him and teach him exclamations and, later, sentences. They teach the child new pain-behaviour. 'So are you saying that the word "pain" really means crying?' —On the contrary: the verbal expression of pain replaces crying, it does not describe it" (*PI*, 244, emphasis original).

'object and name,' the object drops out of consideration as irrelevant" (*PI*, 293). Hence Wittgenstein argues that philosophical problems evolve from treating mental processes and states in the same way as that of speaking about tables, buildings, etc.[11] In a key passage, Wittgenstein depicts this picture of referentialist language:

> How does the philosophical problem about mental processes and states and about behaviourism arise?—The first step is the one that altogether escapes notice. We talk of processes and states, and leave their nature undecided. Sometime perhaps we'll know more about them—we think. But that's just what commits us to a particular way of looking at the matter. For we have a certain conception of what it means to learn to know a process better. (The decisive movement in the conjuring trick has been made, and it was the very one that seemed to us quite innocent.)—And now that analogy which was to make us understand our thoughts falls to pieces. So we have to deny the yet uncomprehended process in the yet unexplored medium. And now it looks as if we had denied mental processes. And naturally we don't want to deny them. (*PI*, 308)
>
> What is your aim in philosophy?—To show the fly the way out of the fly-bottle. (*PI*, 309)

At this juncture, it is natural to ask: How does one reasonably move from simple puzzles over color, pain, and mental processes or states to a critique of philosophy? Answering this question requires the right spirit of reading Wittgenstein (*T*, 27), seeking to discern therapeutic "connections" amidst the pictures and images he paints (RF, 133). Kerr, for instance, addresses language games by pointing to a connection between metaphysics and magic in Wittgenstein's thought.[12] As quoted above, *PI*, 308 speaks of the "conjuring trick" of moving undiscerningly from objects to mental processes, and in *PI*, 109, "philosophy is a struggle against the bewitchment of our understanding by the resources of our language." Hence arises

11. Fogelin, "Wittgenstein's Critique of Philosophy," 43.

12. Kerr, "Metaphysics and Magic," 240–58. According to Rush Rhees's intro to "Remarks on Frazer's Golden Bough," Wittgenstein contemplated including these statement in the text but later edited them out: "I now believe that it would be right to begin my book with remarks about metaphysics as a kind of magic./But in doing this I must not make a case for magic nor may I make fun of it./The depth of magic should be preserved.—/Indeed, here the elimination of magic has itself the character of magic./For, back then, when I began talking about the 'world' (and not about this tree or table), what else did I want but to keep something higher spellbound in my words?" (*RF*, 116–17).

the analogy of metaphysics as a kind of magic that Wittgenstein in part succumbed to in the *Tractatus* by seeking "to keep something higher spellbound in my words" (RF, 116–17; see *PI*, 105, 106). Kerr draws attention to Wittgenstein's comments on Plato's *Theaetetus* (201e–202b) where Socrates speaks of primary elements of which "we and everything else are composed," and there can be no further explanation or account for primary elements other than its name. Wittgenstein aligns both his own tractarian "objects" and Russell's "individuals" to such primary elements (*PI*, 46). The counter to simple, primary elements, however, in the *Investigations* is to question the baseline. What are the simple constituent parts of reality? What are the simple constituent parts of a chair? The wood, molecules, or atoms (*PI*, 47)? The point is that "simple" means "not composite," and hence the baseline is set inevitably through the language game between users of these terms. Missing this feature of language contributes to significant problems of metaphysics:

> [Metaphysics is problematic] in the sense that certain words are allowed to call things "of a higher order" into existence, like an enchanter's spell. "The idea," he says, is "that one can beckon a lifeless object to come, just as one would beckon a person." The principle at work, in metaphysics as in magic, is that of "personification"—animism, as we might say. We have a magical conception of our signs.[13]

The conjurer's trick is perhaps most evident in demonstrable pronouns such as "this" or "that." The natural impulse, it seems, to avoid confusion is to claim that "this" or "that" do not name anything. "Yet, strange to say, the word 'this' has been called the *real* name; so that anything else we call a name was one only in an inexact, approximate sense" (*PI*, 38, emphasis original). Wittgenstein critiques "this odd conception" as a sublimation of the "logic of our language," and he suggests that demonstrative pronouns must be distanced from names. We call "*very different things* 'names,'" and the word "name" summarizes the process of associating terms with those things, but the words "this" or "that" do not align in the same way with names for Wittgenstein:

> It is quite true that in giving an ostensive definition, for instance, we often point to the object named and utter the name. And likewise, in giving an ostensive definition, we utter the word "this" while pointing to a thing. And also, the word "this" and a name often occupy the same position in the context of a

13. Kerr, "Metaphysics and Magic," 248.

> sentence. But it is precisely characteristic of a name that it is explained by means of the demonstrative expression "This is N" (or "That is called 'N'"). But do we also explain "That is called 'this,'" or "This is called 'this'"? (*PI*, 38, emphasis original)

The point is to note a difference in ourselves—the way we view things—when we move across the boundary between naming an object and characterizing an object as a "thing." The former we recognize as *naming*, but do we see the latter as *thinging*? Do we magically invoke "some kind of direct hook-up between thought and thing?" Little seems more powerful or "extraordinary" than giving things names.[14]

> This is connected with the conception of naming as a process that is, so to speak, occult. Naming seems to be a *strange* connection of a word with an object.—And such a strange connection really obtains, particularly when a philosopher tries to fathom *the* relation between name and what is named by staring at an object in front of him and repeating a name, or even the word "this," innumerable times. For philosophical problems arise when language *goes on holiday*. (*PI*, 38, emphasis original)

The phrase "language goes on holiday" is well known, and in context here Wittgenstein's aim is to recall language to its practical "workday" routines amidst relational networks and concrete language games. When on holiday, our language runs the risk of "imagining naming as some remarkable mental act, as it were the baptism of an object" (*PI*, 38). Wittgenstein's metaphor is striking, for Christian baptism is a ritual re-creation with the deepest ontological significance. Likewise, the conjuring-baptism-creating imagery is deeply significant in light of the *Logos* in Genesis 1 and John 1 who *speaks* creation into existence. Here it would be a grave mistake to misread Wittgenstein as promoting a demythological and flattened view of language that proposes grammatical "facts," a grounding for *all* reality within language games. Language games are not modes of creating meaning as much as therapeutic methods for correcting misplaced, presumed "remarkable mental acts." Wittgenstein does not seek to replace one theory of metaphysical meaning with merely another theory via a constructive view of language games (*PI*, 103). We must look elsewhere for a more constructive understanding, and in no way does it seem that Wittgenstein is critiquing magic or ritual as such. The metaphor of magic is employed to critique mentalist meaning, a sort of Kantian idealist[15] approach to

14. Kerr, "Metaphysics and Magic," 249.

15. "Kant's purpose could be described as an ideal language project based on

language: "But in doing this I must not make a case for magic nor may I make fun of it./The depth of magic should be preserved" (RF, 116–17). In this way, we might see Wittgenstein speaking of conjuring, occultish metaphysics as a dark *invocational* magic that calls into existence some*thing* solely from out of the powers of the mind rather than an illumining *incantational* magic that sings within the harmonies of the tangible, existential cosmos. Wittgenstein's targets are cognitivism, scientism, and theorizing, while his acerbic critique of Sir James Frazer's *The Golden Bough* explicitly reveals his openness to the magic of religious ritual.[16]

Turning to our second theme of logical perfectionism, we are not departing to something entirely new, but rather examining a closely related counterpoint in Wittgenstein's critique of language. Referentialism and logical perfectionism (or logical form) are inseparable. In addition to a name representing an object within a proposition (*T*, 3.22), a "proposition is a picture of its state of affairs, only in so far as it is logically articulated" (*T*, 4.032). This is the case for the tractarian Wittgenstein because "in the proposition there must be exactly as many things distinguishable as there are in the state of affairs, which it represents. They must both possess the same logical (mathematical) multiplicity (cf. Hertz's Mechanics, on Dynamic Models)" (*T*, 4.04). For Wittgenstein, both the semantic-representational and the structural-logical constitute the essence of language itself. As Edwards notes, "names name objects" and "configurations of names depict the possible configuration of objects."[17] Hence referentialism and logical perfectionism are two sides of the same coin, and both necessarily constitute propositions, the totality of which is language (*T*, 4.001; see *T*, 5) and even thought (*T*, 4).

As referentialism and logical perfectionism are cojoined, Wittgenstein's critique must address them both, and in logic we can see again the magical misunderstanding of seeking—even demanding the *reality* of—"what" is hidden:

> We want to say that there can't be any vagueness in logic. The idea now absorbs us that the ideal "*must*" occur in reality. At the same time, one doesn't as yet see *how* it occurs there, and doesn't understand the nature of this "must." We think the ideal must be in reality; for we think we already see it there. (*PI*, 101)

---

analysis of propositions or judgments, rather than [a Wittgensteinian] empirical project based on their use" (Garver, "Philosophy as Grammar," 154–53). See also *PI*, 81.

16. See Wittgenstein, "Remarks on Frazer's Golden Bough."

17. Edwards, *Ethics without Philosophy*, 14.

Even beyond demanding crystalline purity in logic (*PI*, 107), the question of logic as a fixed requirement or "normative science" arises (*PI*, 81). Logic seems inevitable and necessarily present to some degree, yet Wittgenstein is not settled on a fixed perfection of logic, as though "what we were talking about in logic were an ideal language" and that only a logician is able "to show people at last what a proper sentence looks like" (*PI*, 81). Again, the analogue centers on the notion of games, and Wittgenstein compares logic in language to rules within a game and words as objects, say, as pieces in chess (*PI*, 108). The comparison is dangerous, however, if it leads us down the path of assuming there exist perfect rules or calculi for games. The tendency then is to think of our language as only an "*approximate*" of an ideal language (*PI*, 81, emphasis original), or even that our actual language may indeed embody such an ideal but only in a way that is mysteriously internal and hidden from us.[18] Either way, we fall into confusion by assuming the "strict and clear rules" of logic are "unshakable," governing absolutely and ideally as a priori realities to be discovered (*PI*, 102–3):

> When we believe that we have to find that order, the ideal, in our actual language, we become dissatisfied with what are ordinarily called "sentences," "words," "signs." The sentence and the word that logic deals with are supposed to be something pure and clear-cut. And now we rack our brains over the nature of the *real* sign.—Is it perhaps the *idea* of the sign? Or the idea at the present moment? (*PI*, 105, emphasis original)

The burden of seeking such crystalline purity with whatever means is at our disposal seems too much: "We feel as if we had to repair a torn spider's web with our fingers" (*PI*, 106). For Fogelin, Wittgenstein's "assault" on the presumed purity, sublimity, or perfection of logic "cuts deeper than his attack on naïve referentialism."[19] Wittgenstein's points on logic are more nuanced, for language is a logical and rule-based activity and much of the issue hangs on our views and expectations with logic. Logic is inevitable, and yet for Wittgenstein logic is not a scientific theory to be discovered but rather more of an active practice. "For the crystalline purity of logic was, of course, not something I *discovered*: it was a requirement" (*PI*, 107, emphasis original; see also *PI*, 112). Philosophy then is an activity, a struggle against the "bewitchment of our understanding by the resources of our language" that is held captive amidst considerations deemed necessarily to be "scientific" (*PI*, 109). The point of this philosophical activity is to diffuse scientific logic, reclaiming the use rather than the abstracted explanation of words:

18. Fogelin, "Wittgenstein's Critique of Philosophy," 49.

19. Fogelin, "Wittgenstein's Critique of Philosophy," 44.

> When philosophers use a word—"knowledge," "being," "object," "I," "proposition/sentence," "name"—and try to grasp the *essence* of the thing, one must always ask oneself: is the word ever actually used in this way in the language in which it is at home?—What *we* do is to bring words back from their metaphysical to their everyday use. (*PI*, 116, emphasis original)

Language in its everyday use will follow rules, but the rules are contingent. In place of some unnamed, ideal essence hidden behind language, language games more accurately depict the activities and phenomenon of our language where we see "a complicated network" of affinities, relationships, and "similarities overlapping and criss-crossing: similarities in the large and in the small" (*PI*, 66). Fogelin describes such rules as working satisfactorily while remaining "indeterminate" in the sense of not conforming to the "logician's demand for rigor."[20] For Wittgenstein, theoretical logicians and philosophers meet confusion by embracing essentialism to the point where even games themselves must have some common essence. Wittgenstein's challenge is "don't think, but look!" at the games to observe differences, relationships, and family resemblance in the activities rather than the so-called essence (*PI*, 66). Or, Wittgenstein ponders, "What does it mean to know what a game is? What does it mean to know it and not be able to say it?" Would this awareness be an unformulated definition that would be recognizable to the mind if it were formulated? Such questions inevitably place the "meaning" of the game outside of the game in abstraction. Hence Wittgenstein counters: "Isn't my knowledge, my concept of a game, completely expressed in the explanations that I could give? That is, in my describing examples of various kinds of game, showing how all sorts of other games can be constructed on the analogy of these, saying that I would hardly call this or that a game, and so on" (*PI*, 75).

Closely tied to indeterminacy, Fogelin points to underdeterminacy and incoherence as features of the rules or logic governing language.[21] Underdetermination is not vagueness, but rather the recognition of the inevitable gaps in our understanding of many things.[22] Similarly, the rules of language may at times present contradictions, as in a game where one rule indicates a move while another rule prohibits the same. Regardless of

20. Fogelin, "Wittgenstein's Critique of Philosophy," 51.

21. Fogelin, "Wittgenstein's Critique of Philosophy," 52–56.

22. *PI*, 80 presents the example of a chair that seems for whatever reason to disappear, reappear, and then disappear again. Do the rules of "chair" address such an illusion? Likewise, matters of personal identity may or may not always fall clearly within determined rules. Within the field of bioethics, the rare and challenging instances of intersex ambiguous genitalia may serve as an instance of underdeterminacy.

indeterminacy, underdeterminacy, or incoherence as subtypes, Wittgenstein's goal is to deflate the myth of logical purity to which language must aim. Scientific definitions are not fixed, per se, but rather exist in flux: "what today counts as an observed concomitant of phenomenon A will tomorrow be used to define 'A'" (*PI*, 79).[23] As another way to describe this, Rhees speaks of "the hardness of the logical 'must'" that stems from talking of what *can* and *cannot* be said rather than what *is* and *is not* said.[24] Hence the phrase "logical impossibility" is problematic because of the associations therein that imply or presuppose some kind of pure system. What then is the system that determines what is or is not logical? This is akin to the child's Sunday school teaser: Can God create a rock so large that he *cannot* lift it? The teaser is entirely different if we were to substitute *can* or *cannot* for *will* or *will not*. The question turns on the assumption of an abstracted system beyond the actual use or practice in language and logic. Rhees suggests that "if you spoke of something as 'logically established' or 'logically certain'—it would mean nothing then to speak of the *principles* of logic as logically certain."[25] To allow the criteria within a logical system or language game to validate some*thing* in the abstract beyond the system is a confusion of language. "A" is a physical object, and yet the statement "there are physical objects" is nonsensical (*OC*, 35):

> "A is a physical object" is a piece of instruction which we give only to someone who doesn't yet understand either what "A" means, or what "physical object" means. Thus it is instruction about the use of words, and "physical object" is a logical concept. (Like colour, quantity . . .) And that is why no such proposition as: "There are physical objects" can be formulated. Yet we encounter such unsuccessful shots at every turn. (*OC*, 36)

In terms of the crystalline purity of logic, Wittgenstein challenges the mentalist framework or theoretical grounding with the question of where logic begins. Is there then no objective truth (*OC*, 108)? Far from it.[26] Moreover, Wittgenstein's aim is not a deconstruction of ultimate mean-

23. As an ironic example, consider *OC*, 106 (written between 1949–51) where Wittgenstein imagines the difficulty of speaking *realistically*—imagine a scientist in the 1940s—with a child who has been told by an adult that the adult had traveled to the moon. "No one has ever been on the moon; the moon is a long way off and it is impossible to climb up there or fly there." Naturally, the child will insist, even pointing to a "way of getting there" unknown to the scientist. What reply can be made to that?

24. Rhees, *Wittgenstein's* On Certainty, 50.

25. Rhees, *Wittgenstein's* On Certainty, 50 (emphasis original).

26. Within Wittgenstein's world, is every statement or thesis doomed to nihilism? "If someone were to advance *theses* in philosophy, it would never be possible to debate

ing or reality, per se, but rather a clarification regarding our faculties of perceiving such meaning. The issue concerns our mistakes with logic and language that lead us to inaugurate magically theoretical realities that are not there. Here, as an example, we may imagine in a Wittgensteinian way some empirical impossibility such as a man flying unaided like a bird, something that contradicts our system of physics and our understanding of gravity. We question such, but what if we meet this reply: "we don't know how unaided, human flight is possible, but those who have experienced such flight know they are flying; and even you can't explain everything." To this, Wittgenstein comments: "we should feel ourselves intellectually very distant from someone who said this" (*OC*, 108). We long to test things—empirically and logically—to seek an autonomous scientific certainty that Wittgenstein maintains our language will not support. Hence language can and should be seen as something of an exercise, a pedagogical opportunity to see the limits of what we know through language, especially when it comes to proposed or theoretical knowledge. Rather than remaining suspended in theory, Wittgenstein's critique of philosophy continually calls us back to embodied, tangible, ways of life.

> What *counts* as its test?—"But is this an adequate test? And if so, must it not be recognizable as such in logic?"—As if giving grounds did not come to an end sometime. But the end is not an ungrounded presupposition: it is an ungrounded way of acting. (OC, 110, emphasis original)

Ethics without (Theoretical) Philosophy

> *Now I'm afraid you haven't really got a hold of my main contention to which the whole business of logical propositions is only corollary. The main point is the theory of what can be expressed* (gesagt) *by propositions, i.e., by language (and, which comes to the same thing, what can be thought) and what cannot be*

then, because everyone would agree to them" (*PI*, 128, emphasis original). Clearly, we must look deeper into Wittgenstein. As Schönbaumsfeld comments regarding *PI*, 128: ". . . this doesn't mean that we are left with nothing—as one might fear—but rather with a change in perspective that will no longer make us succumb to certain kinds of temptation. The result of philosophy, if Kierkegaard and Wittgenstein are correct, is therefore not philosophical propositions, but, as the *Tractatus* famously puts it, 'seeing the world aright' (*T* 6.54)." For the sake of comparison, one might also think of Climacus's reminders in *Concluding Unscientific Postscript* that anything is possible in "pure thought," including foregoing the principle of contradiction (Schönbaumsfeld, *Confusion of the Spheres*, 42, 47).

> *expressed by propositions, but only shown* (gezeigt)*; which I believe is the cardinal problem of philosophy.* —Wittgenstein, writing to Russell, 1919[27]

The aim of Wittgenstein's critique of philosophy is not always easy to discern, and, as Wittgenstein's correspondence with Bertrand Russell intimates, there is much to be read between the lines. For this reason, tracing Wittgenstein's metaphors is paramount, particularly connections such as that between metaphysics and dark magic that reveal something of the depth that Wittgenstein sees in language.[28] It is no secret, for instance, that in the *Tractatus* Wittgenstein saw ethics as his main contention embedded within logic and language. Edwards speaks of logic and ethics as the "twin stars" of the *Tractatus*, and without an understanding of both twin concepts the book lacks coherence.[29] This question of coherence is either a glaring fault in Wittgenstein for assembling such "disparate topics," or it brings to light the problem with Russell's strictly analytical reading. Conceivably, one may attempt to read Wittgenstein solely for insights into theoretical logic and positivism, but to do so is akin to stopping short and focusing exclusively on pedagogical exercises that are intended to lead one further onto something of greater importance and beauty.[30] Simply put, the strictly analytical reading of the *Tractatus* struggles greatly to account for key concluding propositions: "Not how the world is, is the mystical, but that it is" (*T*, 6.44) and "There is indeed the inexpressible. This *shows* itself; it is the mystical" (*T*, 6.522). Surmounting the tractarian propositions, having climbed the ladder, one is more enabled to see the world rightly, and the ladder may (should?) be thrown away (*T*, 6.54).

27. Wittgenstein, *Letters to Russell, Keynes, and Moore*, 71.

28. Indeed, we may speak of the transition from the tractarian methodology to that of the *Investigations* as a movement from seeking to map the span or breadth of language—seeking the propositional and formal *essence* of language (*PI*, 65)—to an examination of the poetic depth in persons manifest in the ambiguities of language. "The problems arising through a misinterpretation of our forms of language have the character of *depth*. They are deep disquietudes; they are deeply rooted in us as the forms of our language, and their significance is as great as the importance of our language.—Let's ask ourselves: why do we feel a grammatical joke to be *deep*? (And that is what the depth of philosophy is.)" (*PI*, 111).

29. Edwards, *Ethics without Philosophy*, 19.

30. To read Wittgenstein only to be immersed in the logical-mathematical puzzles he so enjoyed personally is to be led astray, and he was aware of this propensity and potential shortcoming in his own work as a philosopher-guide: "Wittgenstein himself often felt that he had a bad influence on his students. 'The only seed I am likely to sow', he said, 'is a certain jargon.' People imitated his gestures, adopted his expressions, even wrote philosophy in a way that made use of his techniques—all, it seems, without understanding the point of his work" (Monk, *Ludwig Wittgenstein*, 499).

Once the twin stars are acknowledged as distinct yet cohering, Edwards suggests we are penetrating to the core of the *Tractatus* that paints a picture of a disease in the metaphysical self-understanding of Western philosophy: rationality-as-representation.[31] Spanning broadly from the Greek Sophists to Socrates to Descartes to Kant and more, the capacity for thinking—as conceiving accurate representations of the real—stands arguably as the definition of philosophical self-consciousness in the West. This reductivist understanding identifies the intellect primarily with discursive rationality and leads to a solipsistic anthropology of the modern "self" that Kerr identifies as the central target of Wittgenstein's critique. Wittgenstein's goal through language games is to seek freedom from something akin to the "absolute *conception* of reality."[32] Hence Edwards speaks of the *Tractatus* as essentially a "transcendental critique of the concept of representation" that examines the necessary "conditions of representation" and thereby seeks to reveal the boundaries of thinking itself. One need tread carefully in describing this goal of Wittgenstein; on the one hand, he freely admits the possibility of thinking or discursive rationality in the realm of natural science while limiting discursive rationality where "pregnant silence" is more fitting.[33] At the boundaries of thinking and language Wittgenstein's doctrine of showing arises: "That which mirrors itself in language, language cannot represent. That which expresses itself in language, we cannot express by language. The propositions *show* the logical form of reality. They exhibit it" (*T*, 4.121). In other words, "what *can* be shown *cannot* be said" (*T*, 4.1212, emphasis original). Or, as Wittgenstein later states: "Perhaps what is inexpressible (what I find mysterious and am not able to express) is the background against which whatever I could express has its meaning" (*CV*, 16e).

Out of context, Wittgenstein's comments seem remarkably Kantian, and though perhaps less so in terms of opposing the conceptual and the empirical, the limiting of the bounds of sense is clearly present as argued above. This Kantian maneuver, however, stands to be informed by the context of influences such as Schopenhauer, Tolstoy, and especially Kierkegaard, whose unscientific thinking clears the way for the vital work of apophatic silence.[34] Comparing Kierkegaard with Wittgenstein here is particularly helpful for explicating both Wittgenstein's view of language as therapy and his metaphilosophy. Schönbaumsfeld, for instance, highlights the way Kierkegaard and Wittgenstein regarded philosophy as an activity

31. Edwards, *Ethics without Philosophy*, 19–32.

32. Kerr, *Theology after Wittgenstein*, 26; 3–27 (emphasis added).

33. Edwards, *Ethics without Philosophy*, 12.

34. Edwards, *Ethics without Philosophy*, 30–32.

seeking clarity amidst metaphysical confusions.[35] Both philosophers point to the essential importance of the methodological *how* before the *what* of content and philosophical authorship. The newness and originality of philosophical answers are less important than the way in which one proceeds to frame the questions. Profundity for these philosophers is less a "function of the novelty of the philosophical claims advanced, but is rather a result of the way in which existing problems are addressed."[36] In other words, profundity belongs not to philosophical theories but to *Weltanschauung* (*PI*, 122) and to the accompanying philosophical practices that both enable and emanate from such a perspective. This emphasis on methodology, moreover—philosophy as an activity of clarity and understanding rather than a mode of constructing knowledge—leads naturally to a grounding ethical point of reference. As theoretical philosophy lacks a clear view of knowledge and a sound methodology, Kierkegaard and Wittgenstein turn to the ethical will more than the reasoning mind.

> Tolstoy: a thing's significance (importance) lies in its being something everyone can understand.—That is both true and false. What makes a subject hard to understand—if it's something significant and important—is not that before you can understand it you need to be specially trained in abstruse matters, but the contrast between understanding the subject and what most people want to see. Because of this the very things which are most obvious may become the hardest of all to understand. What has to be overcome is a difficulty having to do with the will, rather than the intellect. (*CV*, 173e)

> One who is ignorant and must be given some knowledge . . . and one who is under delusion that must first be taken away . . . direct communication presupposes that the recipient's ability to receive is entirely in order, but here this is simply not the case—indeed, here a delusion is an obstacle. That means a corrosive must first be used, but this corrosive is the negative, but the negative in connection with communicating is precisely to deceive.[37]

The ethical issue lies not with the intellect but more with the resistance of the will. This leads Wittgenstein to see in "a Copernicus or a Darwin" not

35. Schönbaumsfeld, *Confusion of the Spheres*, 38–59.

36. Schönbaumsfeld, *Confusion of the Spheres*, 39. "Amongst Jews 'genius' is found only in the holy man. Even the greatest of Jewish thinkers is no more than talented. (Myself for instance.) I think there is some truth in my idea that I really only think reproductively (*CV*, 18e)."

37. Kierkegaard, *Point of View*, 53–54.

the discovery of a "true theory but of a fertile new point of view" (*CV*, 18e), and in the same spirit Wittgenstein's own work, seeking to "see the world aright" (*T*, 6.54), is one of "clarification with courage: otherwise it becomes just a clever game" (*CV*, 19e). Both Kierkegaard and Wittgenstein thus turn from theses (*PI*, 128), propositions, and metaphysical speculations to attend to more tangible practices, to "what is *actually* going on, rather than in what the metaphysician hopes to see."[38] Moreover, this turn back to concrete reality (*PI*, 107) is a further connection to ethics as thinking and language are rooted in forms of life (*Lebensformen*) for Wittgenstein and the existence-spheres—aesthetic, ethical, and religious—for Kierkegaard.[39] In Schönbaumsfeld's apt phrasing, the point of philosophy free of scientific thinking and theorizing is to guide persons toward what is needed rather than what is wanted, and, despite the half-century gap between Kierkegaard and Wittgenstein, both were concerned with "scientism" and "system building" amidst a "progress-worshipping, technological culture" that shapes the desires of persons in accord with a diseased *Zeitgeist*. It hardly needs to be stated, moreover, that Kierkegaard and Wittgenstein place

38. Schönbaumsfeld, *Confusion of the Spheres*, 42–43 (emphasis original). Schönbaumsfeld draws attention to the focus both philosophers place upon the philosophical activities that determine and form persons rather than philosophical speculation. Schönbaumsfeld points to similar passages from *Concluding Unscientific Postscript* and *Philosophical Investigations*:

> It is from this side [from the side of existence] that an objection must first be made to modern speculative thought, that it has not a false presupposition but a comic presupposition, occasioned by its having forgotten in a kind of world-historical absentmindedness what it means to be a human being, not what it means to be a human being in general, for even speculators might be swayed to consider that sort of thing, but what it means that we, you and I and he, are human beings, each one on his own (Kierkegaard, *Concluding Unscientific Postscript*, 120).
>
> When philosophers use a word—"knowledge," "being," "object," "I," "proposition/sentence," "name"—and try to grasp the *essence* of the thing, one must always ask oneself: is the word ever actually used in this way in the language in which it is at home?—What *we* try to do is to bring words back from their metaphysical to their everyday use (*PI*, 116, emphasis original).

39. Schönbaumsfeld, *Confusion of the Spheres*, 43. Schönbaumsfeld follows Stanley Cavell in aligning forms of life with existence-spheres, and Sandra Laugier connects Hadot's reading of both ancient philosophy and Wittgenstein to one's way of living: ". . . in Hadot's search for a definition of stoicism and more generally of ancient philosophy the idea appears that philosophical change can be accomplished in daily life, through the understanding of the everyday, the down-to-earth (Hadot often quotes Kierkegaard), the ordinary (as Marcus Aurelius describes the beauties of everyday life). The rehabilitation of daily existence is a new type of exercise, and ordinary language, as in Thoreau and Wittgenstein, is the new space of transformation-conversion demanded by philosophy" (Laugier, "Pierre Hadot as a Reader," 325).

philosophers squarely in the middle of this social critique. Philosophers, above all, face the intellectual temptations and diseases of scientism and theory through metaphysics.[40]

In other words, Kierkegaard and Wittgenstein are at one in their aim of trying to expose the *fraudulence* of metaphysics. Both want to show that it constructs edifices in which no one can dwell using only the appearance of building blocks. But it is not that metaphysical speculation does not agree with common sense and this is why we have to reject it—that is, Kierkegaard and Wittgenstein are not advocating a form of pragmatism, as some commentators have supposed. Rather, what they are trying to show is that metaphysics's pretensions are fantastical and confused by exposing (amongst other things) the speculative philosopher's tendency to disregard practice and context, confuse conceptual with factual questions, and cleave to the false idea of the "God's eye" view of the world.[41]

In light of this critique, Kierkegaard and Wittgenstein pursue philosophy as a clarifying practice tied to the ethical work of self-formation. Philosophy is not a matter of external discourses removed from persons as though one were to participate—or simply observe—an external didactic or "metaphysical system-building." Wittgenstein states: "Working in philosophy—like work in architecture in many respects—is really more a working on oneself. On one's own interpretation. On one's way of seeing things. (And what one expects of them.)" (*CV*, 16e). At the heart of philosophy for Kierkegaard and Wittgenstein lie practices of working on oneself, and hence Schönbaumsfeld argues that both philosophers see themselves as "physicians of the soul who attempt to cure the ailments (confusions) of understanding by presenting a new perspective that will dissolve the philosophical problems."[42]

Moreover, the unique writing styles Kierkegaard and Wittgenstein both employ play important rhetorical roles in philosophical therapeutics.

40. Wittgenstein's engagement and tenure with academia was always hesitant and intentionally interrupted as he doubted even that "a normal human being" could remain honest while being a university professor. He often encouraged his students to seek common professions outside the university where, in Edwards's words, they were less likely to succumb to the "disease of understanding which produces the nonsense of literalization." As a "form of life" particularly ensconced in "metaphysical thinking," the modern university served as "a place of ease and affection, where the self-enclosure of abstract thinking is encouraged, [breeding] the *hubris* that despises or patronizes the common lot of most men and women" (Edwards, *Ethics without Philosophy*, 218, emphasis original).

41. Schönbaumsfeld, *Confusion of the Spheres*, 39, 46.

42. Schönbaumsfeld, *Confusion of the Spheres*, 58, 47–59. See Kierkegaard, *Concluding Unscientific Postscript*, 155.

Kierkegaard's pseudonyms create space to allow for prophetic voices to speak to the writer as well as the reader, and Wittgenstein's *Investigations* explicitly seek to induce in each reader the "trouble" of thinking "thoughts of his own" (*PI*, preface 4e).[43] One of the first to introduce Wittgenstein to the French, Hadot was carefully attuned to the rhetorical dynamic of Wittgenstein's work:

> It is therapeutics that is offered to us. Philosophy is an illness of language . . . The true philosophy will therefore consist in curing itself of philosophy, in making every philosophical problem completely and definitively disappear . . . Wittgenstein continues [from the *Tractatus* to the *Investigations*]. . . to devote himself to the same mission: to bring radical and definitive peace to metaphysical worry. Such a purpose imposes a certain literary genre: the work cannot be the exposition of a system, a doctrine, a philosophy in the traditional sense . . . [*Philosophical Investigations*] wishes to act little by little on our spirit, like a cure, like a medical treatment. The work therefore does not have systematic structure, strictly speaking.[44]

In Hadot's reading of Wittgenstein, philosophy's task is to grasp sense and beauty in the ordinary things of everyday existence. This emphasis on the down-to-earth and the ordinary is wed to language because spiritual exercises are manifest through the "habitual and repeated practice and usage of language including speech (the Socratic dialogue, seen by Hadot as a spiritual exercise), writing (considered as well as an exercise), and reading. In short, spiritual exercises reveal the privileged space of the ethical relationship to language."[45] Language is situated as a clarifying exercise within everyday life, and, far from offering a metaphysical magic, language serves as a therapeutic practice to uncover "the idea of a silent ethics, inscribed in the understanding of the limits of meaning and that of a simple, naïve and ordinary relationship to the world. . . . philosophy is not a theory or a body of doctrines but the clarification of our thoughts."[46] Within this framework, ethics can never exist authentically as a philosophical theory or system of thoughts but functions rather as a philosophical activity and way of life.

43. "Consequently, the indirect form of communication that Kierkegaard and Wittgenstein employ respects the ethical integrity of the reader by giving the 'self-activity of appropriation' pride of place. For it is this which will in the end decide whether what we see in the looking-glass is the image of an ape, or the image of an apostle. The mirror itself is silent on this point" (Schönbaumsfeld, *Confusion of the Spheres*, 59).

44. Hadot, "Wittgenstein," 973.

45. Laugier, "Pierre Hadot as a Reader," 325.

46. Laugier, "Pierre Hadot as a Reader," 326.

Like Edwards, Hadot broadens the implications of Wittgenstein—in line with ancient philosophy grounded in modes of *theoria* and *episteme*—to the history of thought in general, defending, as Laugier describes, "a model of spiritual exercises 'independent of all theory.'" Even more (perhaps most) striking here is the way language exercises—even amidst the immanence and limits of what language can say—can guide one to experience "a sense of wonder before the world" or the "oceanic feeling."[47] Not theory, but everyday life stands to offer a path of formation, and hence the transition from the *Tractatus* to the *Investigations* represents Wittgenstein's growing desire to move from *clarifying* ordinary language to actually *following* poetic language. Rather than seeing logic or language as a "normative science" (*PI*, 81) that leads one into metaphysical certainty and normativity, language is better seen as a formative experience or exercise. Laugier sums up Hadot on this point:

> An undeniable link between the *Tractatus* and the *Investigations* is the immanence of language. It is nonsense to wish for an exterior view or approach. We are always inside a "form of life" in language. Spiritual exercises are not a counter-example to, but a confirmation of, this immanence. They are always explicit, spoken: they are *language exercises*. This is why the idea of spiritual exercise is at least as close to the "second" Wittgenstein as the first, and helps us understand the continuity between the two. In the *Investigations* as in the *Tractatus* there is the aim to de-sublimate philosophy, the desire to bring it down to earth (to 'bring the words home from their metaphysical to their ordinary use,' *PI*, 116). . . For Wittgenstein and Hadot, language is always "ours"; we are born with it. This is an ethical truth, difficult to accept: all words are learned, belong to others, have already been said, and we cannot rise, except in illusion, above or outside.[48]

The "difficulty" of language as spiritual exercise lies in the humility to realize the immanence and contingency of the mortal and human "situation in language," and yet the point of addressing this radical finitude is hardly to suggest an ontology of finitude. Rather, more akin to the Stoics, Hadot's Wittgenstein seeks a form of wisdom in "training for death and for the life whose limits are those of my world."[49] The contingency of language moves beyond the misguided search for the authentic or original meaning in

47. Laugier, "Pierre Hadot as a Reader," 327.

48. Laugier, "Pierre Hadot as a Reader," 327 (emphasis original).

49. Laugier, "Pierre Hadot as a Reader," 327–28.

words to an emphasis and valuation on the language that is present. Hence Laugier states, "there are only other words. Language is there, before me."[50] Hadot's point, then, is to see language as praxis, as methodology, and as an ethical "point of access to the ordinary in our lives." Following language means that language becomes a guide and a way to explore the contingency and particularity in human existence."[51] Language is a social act, a communal practice, and a formative "doing" that is no *mere* game we play, but rather, as James Smith describes, functions as a practice to provide "an account of how we embodied, finite, contingent, dependent creatures make our way in the world."[52] Smith then points to *PI*, 89 where Wittgenstein quotes Augustine's *Confessions*: "What *is* this time? If no one asks me, I know; if I want to explain it to a questioner, I do not know."[53] The practice of admitting such contingency is far from fideism or the abandonment of content.[54] Rather, as a sort of apophatic methodology, the difficult exercises of clarifying language point toward a different and higher knowledge beyond philosophical theory. Wittgenstein hence extrapolates from Augustine's words: "Something that one knows when nobody asks one, but no longer knows when one is asked to explain it, is something that has to be *called to mind*. (And it is obviously something which, for some reason, it is difficult to call to mind.)" (*PI*, 89, emphasis original).

At this juncture, it is fitting to ask if Wittgenstein's critique of (theoretical) philosophy leaves room only for silence. Are the difficult things "called to mind" to be spoken, and is that Wittgenstein's primary contribution to the discourses on and within philosophy? In terms of metaphysical theorizing without roots in any fertile, rough ground (*PI*, 107), Wittgenstein's message is clear that such philosophy is diseased and deceptive. Hubris and confusion are the fruits of such theorizing; Wittgenstein sees this and gives over the whole of his life to the prophetic, deconstructive task of clarity.[55]

50. Laugier, "Pierre Hadot as a Reader," 329.

51. Laugier, "Pierre Hadot as a Reader," 329.

52. Smith, *Who's Afraid of Relativism?*, 59.

53. Augustine, Confessions, 11.14 (emphasis original).

54. "One might be tempted to believe that a sensibility that sees the world's existence as a miracle, and that acknowledges the presence of mystery as such, would despise thinking altogether. On a quick reading such a form of life as I have described might sound like anti-intellectual religious enthusiasm, or a strange variant of Zen. But this is a total misconception. Wittgenstein does not despise thinking; he despises attempts at magic that pass themselves off as thinking" (Edwards, *Ethics without Philosophy*, 247).

55. Sontag, *Wittgenstein and the Mystical*, 131. Sontag argues that Wittgenstein's critique of theoretical philosophy must be understood within the context of Wittgenstein's ascetic life. To grasp Wittgenstein's primary message on simplicity and love, readers must confront Wittgenstein's "life and work" as a whole.

Philosophy is an activity, even a "struggle against the bewitchment of our understanding by the resources of our language" (*PI*, 109) for the simple reason that "philosophy cannot deliver the sort of knowledge that is usually considered as its *raison d'être*."[56] Put in such a light, Wittgenstein's metaphilosophy is the antithesis of a systematic metaphysics or natural theology, and this is perhaps where Wittgenstein is so maddening—even toxic—to many philosophers. Wittgenstein stubbornly refuses to let go of the point that long before we seek to enumerate *what* it is that we know—particularly if we are seeking to name some essence or supposed natural, ideal being—we must attend to the methodological *ways of knowing* that enable or disable any such knowledge.

Hence in the face of philosophy pursued as theoretical science, Wittgenstein commands silence, and yet that silence has everything to do with our inability to proclaim scientific statements about mystical and ultimate reality and is no agnostic gesture toward the mystical itself (*T*, 6.522). Our inability to *say* (rightly) anything does not void a positive, constructive value in *showing*.

> That which mirrors itself in language, language cannot represent.
>
> That which expresses *itself* in language, we cannot express by language.
>
> The propositions *show* the logical form of reality.
>
> They exhibit it. (*T*, 4.121, emphasis original)

From the *Tractatus* through the later writings, this distinction of saying/showing is ever present, and while other formulations, systems, or critiques of philosophy certainly make their normative claims on philosophy, Wittgenstein's legacy is a call to action, practices, and ways of life. We can and should imagine a sort of poetic language, an iconic manifestation where the mystical expresses itself in forms of writing, but for the most part Wittgenstein is only able to gesture and hint at such a breakthrough of transcendence to the immanent. Rather than seeking to address the mystical directly, Wittgenstein points instead to the problem that in Monk's phrasing could only "have an existential, never a theoretical solution." The problem stemmed from a misplaced trust in metaphysical theorizing and it was tragically embodied for Wittgenstein in the scientific and technological "progress" that enabled fire-bombs at Dresden, genocidal concentration camps, and atomic bombs dropped on Japan:

56. Horwich, *Wittgenstein's Metaphilosophy*, 1.

> A picture that intruded upon him he wrote, was of our civilization, "cheaply wrapped in cellophane, and isolated from everything great, from God, as it were." The houses, cars, and other trappings of our environment struck him as "separating man from his origins, from what is lofty and eternal, etc." It was as though life itself was coming to an end, suffocated by the trappings of our industrial age. And, of course, it was futile to expect to alter this course by pointing it out. . . . Yet Wittgenstein continued in his work of undermining the way of thinking that, he thought, lay at the root of the whole disaster.[57]

Above all, Wittgenstein longed for "a change of spirit" that would tear off the cellophane to reveal the "living world." Against the colorlessness of theories—and, as he saw it, cold, "grey wisdom" versus the "passion" of faith (*CV*, 56e)—Wittgenstein yearned for the life and the passion he saw in religious practices. Hence Monk draws attention to key phrases from *Culture and Value* that may help to illumine the whole of Wittgenstein's thinking:

> I believe that one of the things Christianity says is that sound doctrines are all useless. That you have to change your *life*. (Or the *direction* of your life).
>
> It says that wisdom is all cold; and that you can no more use it for setting your life to rights than you can forge iron when it's *cold*.
>
> The point is that a sound doctrine need not *take hold* of you; you can follow it as you would a doctor's prescription.—But here you need something to move you and turn you in a new direction.—(I.e. this is how I understand it.) Once you have been turned round, you must *stay* turned round.
>
> Wisdom is passionless. But faith by contrast is what Kierkegaard calls a *passion*.
>
> Religion is, as it were, the calm bottom of the sea at its deepest point, which remains calm however high the waves on the surface may be. (*CV*, 53e, emphasis original)

Wittgenstein's philosophy is intended as nothing less than an "ethical deed" and an attempt "to incarnate a vision of the healthy human life; the transmission of a moral vision—the attempt to reveal its character and make it potent—is the true burden of all [Wittgenstein's] philosophical

57. Monk, *Ludwig Wittgenstein*, 489–90.

work."[58] As the sign of an arrow is only truly meaningful when one turns direction or acts upon the sign, so, too, is Wittgenstein's philosophy an applied philosophy and "*not* a hocus-pocus that can be performed only by the mind" (*PI*, 454, emphasis original). In the *Tractatus*, a healthy human life means the "sense of life" has been clarified as the limits of thought are acknowledged and such a philosophical act in essence gestures to the ineffable (*T*, 6.52), and in Wittgenstein's later works, *the* "sense of life" is abandoned as a "metaphysical fantasy," or self-referential, magical theorizing. *The* sense of life is never a given. As Kerr notes:

> It is neither objective metaphysical realities (whether forms or atoms) nor subjective states of consciousness (raw feels, mental pictures, innate ideas) but *Lebensformen* that are "the given." What is given is the human world: neither meanings in the head, accessible by introspection, nor essences in the objects around us, yielding to analysis, but the order that human beings establish by their being together.[59]

To be clear, Wittgenstein's call to stability in everyday customs and practices is no naïve claim to a "moral panacea." Returning to particular actions and concrete realities will not eliminate the vices and proclivities of human nature. Competing ideals of character and conflicting modes of life are inevitable, and yet what Wittgenstein's "sound human understanding" pursues is an anthropology free of the diseased trappings of rationality-as-representation that produces a "metaphysical gaze." Hence Edwards notes that "vision becomes the fundamental metaphor for one's relation to reality, with the view *sub specie aeternitatis* as the apotheosis of such seeing; and because the sense modality of vision always incorporates distance between seer and what is seen, reality always lies apart from the thinker."[60] *Lebensformen* and Wittgenstein's whole critique of theory is first and foremost a methodological acknowledgment that every individual's sense of self depends essentially on "being with others. . . being in touch with others" in both a physical and psychological way. Kerr again summarizes:

> Nothing is more fundamental to the whole human enterprise than the community that we create in our natural reactions to one another as they have been cultivated and elaborated in a very contingent historical tradition. "Commanding, questioning,

58. Edwards, *Ethics without Philosophy*, 203, 219.

59. Kerr, *Theology after Wittgenstein*, 69. See also *PI*, 7.

60. Edwards, *Ethics without Philosophy*, 221.

recounting, chatting, are as much a part of our natural history as walking, eating, drinking, playing" (*PI*, 25).[61]

To some, Wittgenstein's philosophy may seem narrow, even reductivist, for refusing to forego its stringent method of critiquing theorizing in philosophy. Throughout the depth and richness of Wittgenstein, the whole of his thought is devoted enormously to this critique of metaphysical theses in philosophy (*PI*, 128). So be it, for in one sense such critics are correct.[62] But if Wittgenstein is right in gesturing—sometimes madly—to the false magic of metaphysics, such a corrective and clarifying claim is profound in its simplicity and powerful in that all modern philosophers must confront it. Wittgenstein himself seemingly was never surprised by negative reactions, nor did he assume that his philosophy would out-debate others. This is due precisely to his lack of content—as content is normally understood in philosophy. Seeking to avoid the trappings of theory himself, his philosophical "content" was the *ordo* of clarity amidst the constant reminder that to understand his works means that one is following as guide the whole life of a person, however imperfect he may be.[63] Despite the limitations of Wittgenstein in going beyond silence or—employing in some way his own understanding of religious thinking—to "soften his own dissolution" enough to pray[64]—what Wittgenstein means for fields of applied philosophy such as bioethics should not be ignored. Wittgenstein's legacy of critique needs to be recovered not as a new antitheory theory but for the way it changes the moral perspective of individuals.[65] Hence even as Wittgenstein himself does not move past silence, his silence is meaningful as it furthers a view of

61. Kerr, *Theology after Wittgenstein* 76.

62. ". . . I was thinking about my philosophical work and saying to myself: 'I destroy, I destroy, I destroy—'" (*CV*, 21e).

63. Monk recounts two of Wittgenstein's students, D. A. T. Gasking and A. C. Jackson, stating the difficulty in understanding Wittgenstein's lectures that contained so may interconnected examples and "repetitive concrete detailed talk." Of this, Wittgenstein commented: "I am showing my pupils details of an immense landscape, which they cannot possibly know their way around" (*CV*, 56e). "In teaching you philosophy I'm like a guide showing you how to find your way round London. . . . Of course a good guide will take you through the more important streets more often than he takes you down side streets; a bad guide will do the opposite. In philosophy I'm a rather bad guide" (as quoted in Monk, *Ludwig Wittgenstein*, 502).

64. "I cannot kneel to pray because it's as though my knees were stiff. I am afraid of dissolution (of my own dissolution), should I become soft" (*CV*, 56e).

65. "One must start out with error and convert it into truth. That is, one must reveal the source of error, otherwise hearing the truth won't do any good. The truth cannot force its way in when something else is occupying its space. To convince someone of the truth, it is not enough to state it, but rather one must find the path from error to truth" (RF, 119).

the world not governed by scientific, rationalistic, technocratic ontologies of power. In a Wittgensteinian way, the lines are drawn between a view of the world as a possession under the guise of rationality-as-representation and a view of the world as miraculously "there":

> Not *how* the world is, is the mystical, but *that* it is.
>
> The contemplation of the world sub specie aeterni is its contemplation as a limited whole.
>
> The feeling of the world as a limited whole is the mystical feeling. (*T*, 6.44 & 6.45, emphasis original)

This position on wonder, moreover, forms the backbone of Wittgenstein's "Lecture on Ethics," a lecture that summarizes numerous points of Wittgenstein's thought on the whole. The experience "*par excellence*" that defines what Wittgenstein means by absolute or ethical value is "best described [by saying that]. . . *I wonder at the existence of the world*" (*LE*, 11, emphasis original). To look upon the world through the theoretical, scientific lens is to see facts and similes; the facts are relative judgments of value—i.e., "this man is a good runner" or "this is the right way to Granchester" (*LE*, 6)—and the similes are the modes of language that carry relative judgments. To say "this is a good fellow" and then note "this is a good football player" seems to imply some common sense of goodness, and yet for Wittgenstein the point we must notice is the use and meaning we gain out of the relationship between the two "goods" as similes of relative judgment:

> In ethical and religious language we seem constantly to be using similes. But a simile must be the simile for something. And if I can describe a fact by means of a simile I must also be able to drop the simile and to describe the facts without it. Now in our case as we try to drop the simile and simply to state the facts which stand behind it, we find that there are no such facts. (*LE*, 14–15)

These statements on wonder draw us beyond seeking facts scientifically to seeing the mysterious world as a miracle. "The truth is that the scientific way of looking at a fact is not the way to look at it as a miracle. For imagine whatever fact you may, it is not in itself miraculous in the absolute sense of the term" (*LE*, 17). For Edwards, these passages on wonder illustrate the "characteristic recognition" of a *Pathos* that is inherent in the existence of something, as well as "a particular response to that *Pathos*." In lieu of scientific hypothesizing regarding facts, one becomes curiously open to a miracle; "the recognition that '*the world is there*' will produce wonder and humility, properly ethical sentiments. . ."[66]

As a way of drawing numerous themes together, it is worth reflecting on what we may identify less as Wittgenstein's constructive but rather redemptive view of language. Namely, while the experience of humility in seeing the world as a miracle—something *there* as a gift, an occasion for love that occludes the will to power—is not conjured through language, language itself in manifold ways bears witness to something greater. "Now I am tempted to say that the right expression in language for the miracle of the existence of the world, though it is not any proposition *in* language, is the existence of language itself" (*LE*, 17, emphasis original). Immediately after suggesting that, however, Wittgenstein admits that some will see the shift from expressing the miraculous "*by means* of language" to the expression "*by the existence* of language" is merely Wittgenstein's way of saying again "that we cannot express what we want to express and that all we *say* about the absolute miraculous remains nonsense." The answer for such pundits will likely seem clear: when such experiences arise that "tempt" us to attribute a quality "which we call absolute or ethical value and importance," what we are seeing is that such words are not indeed nonsense. The experience we are seeking to describe has absolute value just as "*a fact like other facts*," and that the remaining problem is to find the "correct logical analysis of what we mean by our ethical and religious expressions" (*LE*, 18, emphasis original). In other words, has Wittgenstein merely failed through language to speak properly of what he maintains is unsayable?

> Now when this is urged against me I at once see clearly, as it were in a flash of light, not only that no description that I can think of would do to describe what I mean by absolute value, but that I would reject every significant description that anyone would ever suggest, *ab initio*, on the ground of its significance. That is to say: I see now that these nonsensical expressions were not nonsensical because I had not yet found the correct expressions, but that their nonsensicality was their very essence. For all I wanted to do with them was just to go beyond the world and that is *to go beyond* significant language. My whole tendency, and I believe the tendency of all men who ever tried to write or talk Ethics or Religion, was to run against the boundaries of language. This running against the walls of our cage is perfectly, absolutely hopeless. Ethics, so far as it springs from the desire to say something about the ultimate meaning of life, the absolute good, the absolute valuable, can be no science. But it is a

66. Edwards, *Ethics without Philosophy*, 234, (emphasis original).

> document of a tendency in the human mind which I personally cannot help respecting and I would not for my life ridicule it. (*LE*, 17–19, emphasis original)

The passage above is lengthy, yet notably summarizes Wittgenstein and in particular speaks to the question of the relationship between language and the mystical.[67] At one point, in the clear propositions of the *Tractatus*, the mystical lay *beyond* language. After discerning his explorations with language games and *Lebensformen*, however, Wittgenstein's position is more nuanced as he came to see *through* language—the existence of language itself—something of the miraculous and mystical. Yet Wittgenstein hesitates to overstate this. In the passage above, he is "tempted" in saying as carefully as he does that the miraculous is presenced in the existence of language, and amidst this tension it seems we glimpse both the greatest relevance of Wittgenstein to ethics as well as the limit of what he offers. On the one hand, Wittgenstein's constant emphasis on methodology leads to philosophy as an activity of clarity, not a system or paradigm that constructs theory. Wittgenstein's "running against the walls of our cage" is propaedeutic for the revelation of ultimate meaning and absolute value. For an applied field of ethics such as bioethics that exists under the metaphysical gaze of principlism and scientific theories, this insight is subversive to the core. Moreover, the ascetic practice of seeking clarity shifts the focus from an external drive to uncover or possess knowledge as an almost technological object to an internal process that prepares one for, as Kallenberg notes, "the possibility of *ekstasis*" as the seeker is carried up to behold wondrous things far beyond himself, beyond nature. In this way, Wittgenstein sublimates metaphysics (*meta ta physica*) away from theorizing toward *theoria*; after all, Wittgenstein never sought to *refute* the metaphysicians but to *discipline* their use of language. The nature of language "[prevents] direct talk about the mystical" such that the "cardinal problem in philosophy for Wittgenstein was the problem of becoming skilled enough to perceive what can be shown, but not said."[68]

67. Indeed, this relationship between language and the mystical enacts a premodern, theological move against *philosophical* metaphysics as the "positing of reality apart from God" that would seem directly applicable to Cunningham's broader thesis (Cunningham, "Wittgenstein after Theology," 77). Far from supporting an *ad hoc* transcendentalism, Wittgenstein's embodied practices with language as described in the *Philosophical Investigations* and *Lecture on Ethics* stridently contest transcendental and metaphysical theorizing. Pace Cunningham, Wittgenstein works within a neo-Kantian framework while soundly refuting Kantian *a priori* metaphysics (Kerr, *Theology after Wittgenstein*, 37; Cunningham, "Wittgenstein after Theology," 85–86).

68. Kallenberg, *Ethics as Grammar*, 170–71.

On the other hand, for all the critical advantages of Wittgenstein's methodology, by his own recognition, greater meaning remains beyond reach. What remains is the ascetic practice of clarity, the "trouble of thinking" that others must undertake on their own (*PI*, preface 4). Frederick Sontag takes on that trouble, and in seeking to explicate Wittgenstein's gesturing Sontag suggests a "Gospel according to Wittgenstein" where Wittgenstein's lived pursuit of "philosophy, intellect, simplicity, ascetic practice and the search for love" forms a life-work as a "good news." Unlike Tolstoy, who wrestled explicitly (and dangerously) with the Gospels, however, Wittgenstein's "good news" is "intended essentially only for philosophers" as it remains an activity of clarity for thinking.[69] As such, Wittgenstein becomes something of a religious thinker by examining the phenomena of human logic and language that frees him of metaphysical theorizing and of any pretense that language may depart from engagement with the world (*PI*, 90). The mystical that lies no longer *beyond* language may be realized in some way *through* language, and yet this is surely a limited gain. Wittgenstein's deep respect for the thick meaning of rituals (see "Remarks on *Frazer's Golden Bough*") testifies to other modes of knowing and experiencing religious meaning beyond the philosophical practices of clarity. Philosophy for Wittgenstein may be clarified entirely through the analysis of ordinary language, but as Sontag notes: "if we released the fly of perplexity from the fly bottle of intricate metaphysics, we know that Wittgenstein never found release or knew the gift of love. . . . In fact, freedom from his intellect as a taskmaster—this is precisely what he never knew. It was, as he says, at the same time his single talent and his most perplexing demand."[70]

Given the limits of Wittgenstein's method, what can we say the method may accomplish? The answer proposed here is nuanced for good reason. The field of bioethics is ripe for the philosophical practices of clarity that Wittgenstein lived and preached, and such a critique is no small accomplishment for Wittgenstein's method. The method of the *Investigations* considers all language games, and the chapters to follow aim to illustrate how language games bring clarity to controversial issues in contemporary bioethics. The ethical norms and principles employed commonly in the field enact a scientific and metaphysical magic that should be decried. Moreover, employing Wittgenstein's method we find the space for honest engagements of difference not amidst metaphysical conjuring but within the clarity of language rooted within *Lebensformen*. This is the work of shunning disembodied language that floats in abstraction and turning rather to a different

69. Sontag, *Wittgenstein and the Mystical*, 126, 126–31.

70. Sontag, *Wittgenstein and the Mystical*, 129.

conception of how human beings mesh with the world more "instinctively" and ritually as embodied persons. Chapters 4 and 5 hence explore how bioethics should move beyond such abstracted theorizing.

At the same time, while engaging Wittgenstein's method with bioethics, it would be disingenuous to assume that the limits of gesturing toward absolute value can solve the perplexities of applied ethics. In this regard, as Sontag comments, the method of Wittgenstein's later writings "shuts nothing from being voiced in words" and raises the question: Can his method "take away nothing, leave everything as it was, except remove our perplexity, bring us into agreement and make further speculation in philosophy unnecessary?"[71] For all the value in practices of clarity, at the end of the pursuit, how does one find rest? Or, to follow Wittgenstein's strong metaphor, might there be—and where is—a book on ethics that would, with an explosion, destroy all other books in the world (*LE*, 9)? At the risk of looking too far ahead, the final chapter of this book steps into this question of ethics, ultimate value, and the ultimate good.

71. Sontag, *Wittgenstein and the Mystical*, 129.

4

Bioethics as a Self-Referential Language Game

Identifying Scientific Bioethics: Transscience and Transethics

The inconsistency of a science professing that it can explain all human action without making value judgments, while the scientist's private actions are said to be often motivated by moral motives, can be more simply demonstrated the other way round. If the social scientist can explain all human actions by value-free observations, then none of his own actions can claim to be motivated by moral values. Either he exempts himself from his own theory of human motivation, or he must conclude that all reference to moral values—or any values—are meaningless: are empty sounds.[1]

FOR WITTGENSTEIN, THE AMBITION of scientific theorizing is no small thing. The aim of scientific theorizing is "to solve its problems by bringing out into the open what once was hidden. (The 'analysis of concepts' is supposed to be a primary means to this end.)" The assumption within this process is that philosophical and ethical problems are inherently "logical" and hence are "solved by the discovery of objective and eternal philosophi-

1. Polanyi, *Knowing and Being*, 34.

cal (logical, conceptual) truths."[2] These epistemological assumptions are visible within modern ethics and bioethics, and the *modus operandi* of scientific theorizing in bioethics is to assume some ethical content that is held in common—i.e., common morality—and then apply those principles through a moral calculation (e.g., Beauchamp and Childress, *Principles of Biomedical Ethics*, 8th ed.). Others in bioethics employ theoretical methods such as utilitarianism, deontology, feminism, professionalism, etc., and yet both common morality and theories in bioethics share a similar status of being suspended amidst propositions disconnected from any explicit ontological or theological grounding. As Anscombe, MacIntyre, Hauerwas, Engelhardt, and others have articulated, ethics must first make sense of moral foundations before pressing on with any applied ethics. Bioethics predominately ignores this, hence the relevance of Wittgenstein's call to move past scientific thinking to engage ethics more authentically.

Moreover, the parallels between Wittgenstein's scientific thinking and secular thinking are not difficult to see. Wittgenstein's use of "scientific thinking" is technical in the sense of extending beyond observations with nature to the moral and ethical sphere of human life, and in so far as secularism seeks a generalized view from nowhere, scientific theorizing assumes in some sense an ethical objectivity and universality apart from history and narrative. In tracing Wittgenstein's move beyond this scientific theorizing, we come to a religious form of thinking that stands to change drastically the form of bioethics.[3] Wittgenstein's religious thinking is the focus of the final chapter, but for now it is helpful to examine a further element of scientific thinking on natural science that pertains to scientific thinking on ethics.

Scientific thinking for Wittgenstein entails a broad "analysis of concepts," and many today no doubt will find it difficult to accept Wittgenstein's description of scientific thinking as it runs contrary to the spirit of the modern world that surrounds us on all sides. More likely, they would agree with

2. Edwards, *Ethics without Philosophy*, 112. Edwards further describes three key features of the "scientific model" of philosophy: 1) this model "stresses the explanatory aim of philosophy" and "attempts to explain particular facts or problems by the construction of general theories or analyses"; 2) the scientific model emphasizes discursive argumentation as the way through which "philosophical truths are discovered and made known"; and 3) this model of philosophy naturally appraises philosophical claims in terms of "truth and falsity, plausibility and implausibility, correctness and error" (112–13).

3. This change in the form of ethics and bioethics does not entail an immediate claim of a particular religious or theological tradition; such a claim, moreover, would not be fitting with Wittgenstein's sincere and longstanding—yet generalized and noncommittal—interest in religion. With Wittgenstein, it seems fitting in a therapeutic way to first note the need for religious thinking before entering discussions of *which* religion or *what* theological particularity is at hand.

the social expectation that science primarily exists and is capable to remedy conflicts that arise between the natural world and society. Science is seen as the agent responsible for addressing practical or real-world problems, and then society itself will decide politically how or whether to act on the available scientific knowledge. Science produces some ability, and politics does or does not act on that ability. The scientist describes if it is possible to withdraw oil in the Antarctic while politicians decide if that is acceptable. Such a generalization of science and politics, as nuclear physicist Alvin Weinberg (1915–2006) notes, is helpful and yet "oversimplifies" the line between means and ends in that such an example is not exact:

> What is thought to be a political or social end turns out to have numerous repercussions, the analysis of which must also be assessed in moral and political terms; or what is thought to be a scientific means has non-scientific implications which also must be assessed in these terms. The relationship between the scientist and the politician is thus far more complicated than the simple model described [here].[4]

That said, the tension or space between science and politics is helpful for making an important point that moves beyond simplistic explanations for science and politics, respectively: namely, "many of the issues which arise in the course of the interaction between science or technology and society—e.g., the deleterious side effects of technology, or the attempts to deal with social problems through the procedures of science—hang on the answers to questions which can be asked of science and yet *which cannot be answered by science*."[5] Weinberg labels these questions that cannot be answered by science "*trans-scientific*" questions for these questions epistemologically transcend science. These questions involve facts and data and can be stated in the "language of science," and yet by the nature of science these questions move beyond the possible answers science may offer. As an example of this, Weinberg considers the biological effects of low-level radiation insults on mice that scientists can correlate to acceptable levels of radiation exposure for the environment. The experiments and calculations on the scale of mice produce a probable scale or ratio for safe radiation exposure, and yet the assumption remains that the safe exposure for mice will correlate to a safe exposure rate when expanded greatly. In

4. Weinberg, "Science and Trans-Science," 210. As the leader of the Oak Ridge National Laboratory (TN) and participant in the Manhattan Project, Weinberg was well occasioned to contemplate the boundaries of scientific knowledge and moral/political action.

5. Weinberg, "Science and Trans-Science," 210 (emphasis original).

Weinberg's calculation, however, to adequately test this scientific correlation, 8,000,000,000 mice would be needed, hence rendering the question beyond complete scientific investigation. Similarly, calculating a probability equation for extremely improbable events such as the failure of the Colorado River Hoover Dam requires an immense series of assumptions regarding statistical possibilities for errors. The total probabilities and conditions—including testing ion chambers, transistors, control rod bearings, etc.—are so vast that *testing* and *observing* the conditions for failure are scientifically impossible. These observations are not to devalue the science behind such examples but rather to point out the inevitable trajectory from "pure science" to a transscientific thinking as the meaningful possibilities and implications for humanity rapidly spill over scientific equations.

Weinberg continues by describing the whole of engineering as trans-science, as there can be no exact testing or replica of the Golden Gate Bridge before it is deemed structurally sound. One may test smaller-scale projects such as a jet engine, but the discernment involved with large-scale engineering is technically known as "engineering judgment," and this prudence or judgment trends beyond crystalline scientific knowledge more into Wittgenstein's language games. Another way to see this is to note the line between what is "researchable" and what is "not researchable," and how something is deemed scientific based on that answer. Is the contemporary fight against the Western obesity "epidemic," for instance, something to be handled by science, and if so, what scientific or medical interventions will resolve obesity? Where does science begin and end when we view individuals and societies as willful, desiring, and emotive human beings? Consider, for example, Nobel Peace Prize laureate Norman Borlaug—who has been called the father of the 1960s Green Revolution for his work in the agriculture and biotechnology that saw cereal production rise exponentially—offering a possible answer to the global hunger "crisis." Man's "war against hunger" is often framed as a biotechnological problem, and debates arise between enthusiastic agriculturalists and conserving environmentalists over what constitutes "sustainable agriculture." Yet how much of this issue is "purely" scientific? In the twentieth century, the world population increased three times more than in any previously known era, spiking from 1.5 to 6.1 billion.[6] Can science create a global food "model" when such a model must admit enormous fluidity in the factors affecting world population?[7] Pure science cannot address the vari-

6. Roser et al., "World Population Growth."

7. Borlaug testifies to the phenomenon of trans-science in terms of global hunger: "Most certainly, agricultural scientists have a moral obligation to warn the political, educational, and religious leaders about the magnitude and seriousness of the arable land, food, population and environmental problems that lie ahead. These problems

ables in balancing future population levels as (nuclear) wars, (undiscovered) pestilence, and (genetic monoculture) famines always remain as possibilities. Historically these factors have been great levelers for the human population, and contemporary Westerners in particular experience great stability and peace when measured against other times, but assuming a perpetual stasis seems the height of hubris and presumption. What is clear is that one must either accept the inherent limits of scientific knowing *qua* science, or one must accept with faith what science indicates and promises but cannot quantify on its own terms.

Additionally, another important example of transscience may be seen in choices or values based on "priorities" within science. This was an important matter for Michele Polanyi, who greatly cared for selection bias in scientific research. In the midst of a purportedly neutral and objective search for knowledge, how do we understand the aesthetic or social element of scientific interest or taste in research? How does this tacit element of human persons affect science when science is purported to be a neutral, objective endeavor?[8] For Weinberg, this scientific choice runs parallel to the different styles of science—e.g., "pure versus applied, general versus particular, spectroscopy versus paradigm-breaking, search versus codification"—for these "scientific values" reflect "taste rather than scientific truth." Additionally, this transscientific thinking can be seen as social priorities are established through scientific elements. In the case of *Brown vs. Board* (1954), for instance, Weinberg notes that the Supreme Court called upon a "scientific doctrine" to claim that "educational deprivation caused psychological damage to the individual as an important argument for its decision to order desegregation."[9] Far from neutral or value-free, as with all human endeavors, science is inherently biased and historically conditioned.

In addition to reminding us of the very personal and human borders around scientific thinking, transscience aligns well with Polanyi's "republic of science," an establishment presence for science that only accepts as truth the scientific results from those with proper credentials. The boundaries of respectability within the republic are stringent, and a "corpus of science" is established by the predetermined rules:

> [The] function of professional standards in science is but the logical outcome of the belief that scientific truth is an aspect of reality and that the orthodoxy of science is taught as a guide

will not vanish by themselves. Unless they are addressed in a forthright manner future solutions will be more difficult to achieve" (Borlaug, "Green Revolution Revisited," 21).

8. Polanyi, *Personal Knowledge*.

9. Weinberg, "Science and Trans-Science," 214.

> that should enable the novice eventually to make his own contacts with this reality. The authority of scientific standards is thus exercised for the very purpose of providing those guided by it with independent grounds for opposing it. The capacity to renew itself by evoking and assimilating opposition to itself appears to be logically inherent in the sources of the authority wielded by scientific orthodoxy.[10]

In essence, a sort of a language game is played as (1) science is proclaimed valid by virtue of being credentialed or "co-ordained" within the scientific, "coordinating" community, and (2) this community is distinct as an "elite" sector within the larger society. Within these rules of the republic of science, only a highly credentialed scientist may speak on the qualities of radioactive wastes, but within the broader "republic of society," for instance, any citizen—such as property owners near the waste site—may speak with vested interests. Again, transscience emerges clearly in such an instance, and this leads to Weinberg's call for honesty and transparency regarding these boundaries. The dynamic line between science and transscience is of utmost importance to identify not for the sake of claiming one can create a perfect model for navigating the immediate space on both sides of the line but rather for the sake of deflating the image of science as a source of truth outside quantified observations in the natural world. As scientists such as Polanyi and Weinberg attest, the nature of science itself leads one to the edges of science, and we are compelled to a transscientific mode of questioning that at the very least must be identified outside the "crystalline purity" of scientific thinking (*PI*, 107 & 108) that so easily extends to modern philosophy as well as to natural science.

What does this transscientific thinking then entail for bioethics? To answer this we must remember the different ethical postures that inevitably arise from different conceptions of how man relates to nature. For example, in a hierarchical, Aristotelian view of nature, we can discern a teleology whereby the nature of the world entails a just order of things in the world. Aristotle's view of nature assumes a perpetual movement from potentiality to actuality, and this change is "intelligible" in a theological way as *Metaphysics Lambda* makes clear. Hence Bouyer notes that "becoming"—understood as the move from potentiality to actuality—assumes "the presence in matter of its future form," and yet the form is not "already" in matter as the process of "becoming" is by definition dynamic. Equally, however, the forms are not existing "independently" and so are present "in a cosmic intelligence, visualized as 'thinking thought' (*noesis noeseos*), which Aristotle considers the

10. Polanyi, *Personal Knowledge*, 6.

prime mover (itself motionless) of the universe."[11] In other words, within an Aristotelian view of the world, it is fitting to consider ethics as a practice of living *within* the cosmos for the cosmos is orderly.[12] The relationship between science and ethics is harmonious, and while the modes of knowing will change depending on the scientific or moral matter being pondered, the method and end will be theological as science and ethics alike are cooperative within an orderly, teleological cosmos.

By strong contrast, particularly visible in the last 150 years, the predominant *mythos* and scientific narrative now denies "nature as a hierarchy" in favor of "nature in a taxonomic schema." Within such a neo-Darwinian materialist world, the order or scale of nature is not fixed but rather exists more as a "vast and complex family-tree," and to trace an ethical "ought" from within this flux one is almost inevitably left within the wake of Julian Huxley's materialist-evolutionary synthesis of ethics and eugenics. "What Nature is, and how it has reached its present state, [Huxley] tells us as a biologist; and from there he goes on in his capacity as a prophet, to tell us both where Nature is going and more . . . that it is our job to help her along."[13] Within an Aristotelian or teleological framework, the transscientific border would lead a bioethicist to seek harmony, perhaps in the spirit of Ravaisson or Jahr; the whole of biological nature is animated and imbued from within by a meaningful ethical purpose. Within the new mythos of flux, when one moves beyond the borders of science, the result is a turn to dialectic process and the will to power.

Transscience draws out that science is inherently limited in what it may suggest *ought* to be done in light of what it observes is true in nature, and

11. Bouyer, *Cosmos*, 80.

12. Such a vision of harmony between ethics and the world can be seen respectively in Jahr (*Bio-Ethik*) and Potter's ("Bioethics") articulations of bioethics as opposed to the technical emphasis for Hellegers (see Jonsen, *Birth of Bioethics*, 22).

13. Toulmin, *Return to Cosmology*, 56. Toulmin goes on to quote Huxley at length to demonstrate how he progresses from science to ethics: "When we look at evolution as a whole, we find, among the many directions which it has taken, one which is characterized by introducing the evolving world-stuff to progressively higher levels of organization and so to new possibilities of being, action, and experience. This direction has culminated in the attainment of a state where the world-stuff (now moulded into human shape) finds that it experiences some of the new possibilities as having values in and form themselves; and further that among these it assigns higher and lower degrees of value, the higher values being those which are more intrinsically or more permanently satisfying, or involve a greater degree of perfection . . . we can say that this is the *most desirable direction* of evolution, and accordingly that our ethical standards must fit into its dynamic framework. In other words, it is ethically right to aim at whatever will promote the increasingly full realization of increasingly higher values" (Toulmin, *Return to Cosmology*, 59, emphasis original).

one's view of nature is of the greatest importance in this equation. Moreover, this brings us to a critical point: in a real sense, bioethics can obscure the appropriate limits within Weinberg's transscience when bioethics is adopted as a confirming "moral stamp" upon some human subjects research, clinical trial, or other such scientific activity. Bioethics that ignores biopolitics veils the implication of Polanyi's "republic of science" that science is hardly as neutral and objective as many expect. In other words, the chief danger Weinberg and Polanyi both identify is to expect or ask too much from science, and then the stakes are raised when bioethics helps cover over the transition from what science can and cannot say.

Ironically, amidst this debate on how to frame our approach to nature, it can seem natural to follow a scientific mode of thinking with natural science into a similar scientific mode of thinking with ethics—seeking analysis of ethical *concepts*, pursuing ethical matters as *logical problems* to be solved. This is precisely a methodological extension from modern science into modern ethics. Post-Baconian modern science seeks technical power by discovering objective and eternal physical truths, and bioethics can enact the same hubris by seeking moral, universal principles.[14] Weinberg's transscience hence can serve as a significant parallel representation for scientific thinking with morality and bioethics, and the problem of bioethics expertise discussed in chapter 1 is a perpetual reminder that something is off. The presumed "clarity of science" view is flawed, and the presumed clarity of ethical transcendentals or intuitions equally falls short. How does one justify expertise in moral theories when the foundation for these theories is unknown? Weinberg quotes the late professor Harold Laski to make a similar point:

> Special knowledge and the highly trained mind produce their own limitations . . . Expertise sacrifices the insight of common sense to intensity of experience . . . It has a certain cast-spirit about it, so that experts tend to neglect all evidence which does not . . . belong to their own ranks . . . [and] where human problems are concerned, the expert fails to see that every judgment

14. This view assumes that (1) the fundamental reality of the world is comprised of material objects, and (2) bioethics then applies the methods of materialism as the way to approach morality as though moral "truths" correlate to material "objects." The irony is that a materialist, objectivist view of the world hardly supports inherently any understanding of moral values. The animal kingdom within nature exhibits a law of instinct and factual realities whereby some dominant males eat their young offspring, and survival is an absolute life *reality*, not a *value* that may undergird beauty, goodness, humility, and mercy.

> he makes not purely factual in nature brings with it a scheme of values which has no special validity about it.[15]

The primary point is that the scientist can have no "monopoly" on wisdom or moral knowledge amidst the transscience maneuver, and in the same way, we may question how the scientific moralist can justify any claim to ethical expertise when the grounding of bioethics remains within scientific theorizing. Anne Maclean's critique of utilitarianism as the normative methodology within bioethics takes up this point, again raising the question of moral expertise in bioethics. In tune with Wittgenstein, Maclean sees the primary contribution of philosophy to be "clarification of issues, types of issue, assumptions, arguments, and concepts."[16] Beyond these more humble limits, utilitarianism follows Mill's proposed *ordo* that generalized theory must precede particular truths, and generalized theory in this case is the grounding upon which the ensuing rational steps are applied. How are those steps reasonable? When they execute the moral logic of the general theory, the system is deemed rational, and the ambition of bioethics parallels the ambition of technomedicine; namely, for the latter, medicine aims to answer and remedy the problems of illness while bioethics purports to answer and remedy the problems of morality. Moral rationality becomes a calculus, a logical arrangement based on the general theory at hand. However, Maclean bluntly explains the real import of what lies within the general theory of utilitarianism: "*what people want* is the ultimate measure of right and wrong; this is the essence of utilitarianism."[17] In short, scientific theorizing applied to ethical thinking supplies a generalized foundation as the starting point that enacts a logical process that is far from any transcendent morality. Again, within this paradigm, ethics is enslaved to scientific thinking, to logical processes, to the whimsical immanence of *what people want*.[18]

Similarly, Charles Foster takes on another touchstone of bioethics, autonomy, in a way that further reveals a thin ethical calculus prominent within bioethics. While noting at least four senses of autonomy—Kantian, psychological, rights-based "constraint on action," and evaluative

15. Weinberg, "Science and Trans-Science," 222.

16. Maclean, *Elimination of Morality*, 203.

17. Maclean, *Elimination of Morality*, 10 (emphasis original).

18. On the matter of utilitarianism and scientific modernity, Yannaras sees utilitarianism as an inevitable extension of modern ethics when a divine "authority" for ethics is rejected. In place of a transcendent, religious ethics, society accepts "on utilitarian grounds, ideas about good and evil which may either be those put forward in each particular case, or those hallowed by experience. And we constantly seek to improve them, using philosophy or science to study the manifestations of social behavior" (Yannaras, *Relational Ontology*, 14).

sense[19]—what emerges primarily from autonomy is a serviceable "device used to further whatever universal law is favoured by its advocates."[20] Far from a secure ethical foundation, the language games surrounding autonomy flex tremendously, and, as a canonical principle of bioethics, autonomy is all but universally unquestioned within the West. Hence Foster ponders a series of questions such as "Whose autonomy?" when it comes to the competing autonomies between a foetus and a mother regarding abortion, or does autonomy bequeath a "right" for incarcerated prisoners to have children, or who possesses autonomy in the Jodie and Mary cojoined twins case?[21] These are typical bioethics questions, and how striking that the fundamental bioethics touchstone of autonomy can in no meaningful way stand alone to provide answers. Far from a neutral, rational discourse seeking concrete morality, Foster's autonomy is a Trojan horse that imparts a form of Kant's Universal Law as the *de facto* morality. What we see then is

> a new form of Kantism in medical ethics, but with a sort of dogmatic, hectoring liberalism replacing Christian morality as the Universal Law. One is only regarded as truly free insofar as one's opinions and one's choices are in line with that Law.[22]

What emerges from this is an *ordo* or progression in scientific ethics that moves from a theory to a rationalized ethics, and perhaps the greatest danger in this ordo is the common presumption of scientific neutrality. For Sherrard, the notion that contemporary scientific theories are "neutral" or "value-free" or "objective" enacts an "insidious fallacy," even a "total lie." He adds:

> The very nature of human thought is such that it cannot operate independently of value-judgements, assumptions and dogmas. Even the assertion that it can constitutes a value-judgement and implies a whole philosophy, whether we are aware of it or not.[23]

Within both science and ethics, the framework for these value-judgements take us to questions of the ultimate good, and Julian Savulescu is a compelling figure for just such an exploration into the inevitable, implicit theology within scientific thinking.

19. Foster, *Choosing Life, Choosing Death*, 7–9.
20. Foster, *Choosing Life, Choosing Death*, 11.
21. Foster, *Choosing Life, Choosing Death*, 21–28, 37–39, 143–61.
22. Foster, *Choosing Life, Choosing Death*, 182.
23. Sherrard, *Human Image, World Image*, 6–7.

Bioethics Denying Scientific Progress—And Good Philosophy?

A means can be justified by its end. But the end in its turn needs to be justified. —Leon Trotsky[24]

By his own admission, Savulescu paints a gloomy picture whereby we twentieth- and twenty-first-century humans are too smart for our own good. The scientific and technological advances of the modern world have made immediate, global annihilation possible. For Savulescu and his frequent coauthor Ingmar Persson, "Ultimate Harm" is possible, even likely, from a number of fronts, including: (1) the possibility of intentional misuse of science, (2) omission of aid to the developing world, and/or (3) failure to cooperate globally, leading to the further collapse of the environment.[25] As a result of this problem, Persson and Savulescu argue five primary points that summarize the predicament of technologically-empowered humans who are endowed with a flawed moral psychology:

1. It is much easier, as a rule, to harm than to benefit.
2. As scientific advances increasingly progress—sped further by cognitive enhancements—so increases the possibility for mass destruction from a few.
3. Even if those immoral enough to cause mass destruction are few in number, it is likely that those immoral few live within huge human populations, with the ability to effect great harm.
4. Moral enhancement of such magnitude to counteract mass destruction is not scientifically possible currently, nor in the near future.
5. Therefore, "the progress of science is in one respect for the worse by making likelier the misuse of ever more effective weapons of mass destruction, and this badness is increased if scientific progress is speeded up by cognitive enhancement, until effective means of moral enhancement are found and applied."[26]

What lies within Persson and Savulescu's gloomy picture is a certain evolutionary hypothesis or anthropological narrative. They assume that human moral psychology is currently adapted for life within technologically limited, small communes where the consequences for disaster exist on a small

24. Trotsky, *Their Morals and Ours*, 34.

25. Persson and Savulescu, "Moral Transhumanism," 662–65.

26. Persson and Savulescu, "Perils of Cognitive Enhancement," 174.

scale. This is a simple, yet enormously critical assumption that the authors unfortunately do not develop at length but rather assert as a touchstone. Fundamentally, it seems, nature has not equipped our moral psychology for life after the mid-twentieth century when technological and scientific progress surpassed human capacities for morality. The scales have tipped, and ironically the only way past this modern problem is to look for a solution further within the very scientific, technological powers that are key factors inciting the problem.[27] In short, Persson and Savulescu's narrative depends upon an assumption of inadequate biology or genetics, a materialist viewpoint where evolutionary progress is lagging or has failed in the face of the remarkable powers of modern *techne*. Embracing this *techne* through biomedical means of moral enhancement may not necessarily be *the* cosmos-saving agent,[28] but the authors make this their focus over familial, economic, and educational reforms to make a point: "many people reject [biomedical enhancements] out of hand for untenable reasons."[29]

This last sentence reveals a notable element in Savulescu's work. As he argues for headline-grabbing, controversial conclusions, he frequently challenges and critiques the field of bioethics for failing to follow "good" (his) philosophical analysis and argument. What is needed, Savulescu argues, is the insight to move beyond a "slavish moralism" characteristic of bioethics. In Savulescu's account, confusions in bioethics arise when Hume's "fact-value" or "is-ought" distinction is ignored at the intersection of science and ethics. Science and ethics are entirely disparate for Savulescu, and to think that the natural facts of science will answer questions of how we should live confuses the distinction between (scientific) facts and (ethical) values:

> The tendency today is to roll over and 'scientify' everything. Evidence will tell us what to do, people believe. But what constitutes sufficient evidence is an ethical decision when we make up our

27. Persson and Savulescu, *Unfit for the Future*, 10–11.

28. Persson and Savulescu do not attempt to identify exactly to what degree biomedical moral enhancements are necessary. There is an "imperative" that scientific research explore "moral bioenhancement" as a "compliment" to traditional means of moral enhancement such as education. Beyond research, however, they argue less for a specific policy of what types of biomedical enhancement must be done when and to whom. Rather, they paint in broad strokes to argue *for* neglected possibilities with biomedical enhancement, whatever those may be. When they do speak of the balance amongst enhancements, they imagine "an interplay between biomedical and social/political techniques rather than the former alone being in the driver's seat" (Persson and Savulescu, *Unfit for the Future*, 124). What seems clear is that humans possess a duty to enhance, but the means should involve some nexus of traditional and biomedical forces.

29. Persson and Savulescu, *Unfit for the Future*, 11.

> minds about what to do. What level of blood pressure, cholesterol and glucose is safe, or healthy, is like what the speed limit or blood alcohol ought to be. It is an ethical judgment about weighing risk and benefit.[30]

What Savulescu does next is to push this distinction between fact and value upon humanists such as Leon Kass, Michael Sandel, and Jürgen Habermas, who object to genetic selection and cloning. The core of their objection centers on generations and the concern of "genetic despotism" (Kass) from one generation to the next that effects a "hyperparenting" (Sandel) and removes the child's freedom when his or her desires for life may not accord with the parents' expectations.[31] For Savulescu, this line of reasoning follows a conceptual confusion. How can one coerce another who has yet to exist? An exception may arise when a parental action might entail that the child's existence is a "life not worth living," and yet, for Savulescu, it is *de facto* senseless to deny the "metaphysical fact" that offspring who owe their existence to their genetic/biological parents might lose autonomy and freedom due to the parents' actions. Kass is well known for aligning cloning with "the wisdom of repugnance,"[32] where following one's moral intuition serves as an ethical compass. Not surprisingly, in some ethics circles Kass's phrase was soon discounted and labeled the "yuck factor" precisely because those such as Savulescu only value quantifiable, utilitarian measures of benefit and harm. Intuition or conscience, for Savulescu, means little except to hold back the trajectory and promise of true science.

Kass, however, sees a meaningful connection amidst generational relationships that posits a value scale beyond materialism or mere biology. On his scale of value, the actions of parents *do* affect genetically modified children who are yet to exist biologically. But this is precisely the sort of value scale that Savulescu critiques in bioethics. Kass speaks of "repugnance" as a *natural* reaction when facing the factual genetics scenario of human cloning, but Savulescu gains some high ground in the argument by simply denying Kass's description of "natural."

Kass's "repugnance" is not the only one of Savulescu's targets. Appeals in bioethics to human dignity mean nothing on his utilitarian scale. Likewise, gaining the consent of human subjects of research for the sake of protecting privacy and confidentiality amounts for Savulescu to little more than a "lethal and widespread malaise."[33] Why? These principles possess

30. Savulescu, "Bioethics," 30.

31. Savulescu, "Bioethics," 28.

32. Kass, "Wisdom of Repugnance."

33. Savulescu, "Bioethics," 30.

no inherent value but rather are subject to analysis against the possibility of massively enhancing knowledge to save and improve lives. What is the greater benefit? To save one's privacy or to save a life? Within this logic, we can even see how Savulescu takes head-on the rhetorical minefield of Nazi eugenics. Savulescu openly favors eugenics and denies that associating eugenics with "this is what the Nazis did" possesses any legitimate criticism. After all, Savulescu argues that our culture's liberal social acceptance of testing for Down syndrome, Fragile X, systic fibrosis, etc. *is* eugenics, and so again, he critiques shallow bioethical arguments by positing a materialist, utilitarian scale of greatest harm/benefit to define what ought or ought not to be socially accepted.

What Savulescu is primarily addressing is the "faith" that bioethics places in mid-level principles as proposed by Beauchamp and Childress.[34] To be sure, there are other approaches to bioethics, but appeals to "dignity," "autonomy," and even "consent" constitute the majority approach that Savulescu sees in the "winning moralists" who are simply "philosophically naïve."[35] In place of "good philosophy," bioethics is bound to principles-based moralism and functions in a religious way rather than via philosophy and sound argumentation.[36]

According to Savulescu, bioethics has failed, and he does not hesitate to ridicule this failure.[37] Moreover, without any more substantial grounding for principles-based bioethics, Savulescu's utilitarian critique proves an earnest challenge for bioethics not unlike Wittgenstein's critique of theorizing as "a kind of general disease of thinking which always looks for (and finds) what would be called a mental state from which all our acts spring as from a reservoir" (*BB*, 143). This is a challenge that reveals how mid-level principles can only be useful when there is some shared moral vision or understanding of the good within which the principles may be defined.[38] Hence Savulescu sees bioethics as "scientifying" ethics, ignoring the distinction between scientific fact and ethical value, and this failure enslaves bioethics to moralism when Savulescu instead pines for an ethics that embraces the open possibilities in a "true" science and medicine that favors calculations for enhancement.

34. Beauchamp and Childress, *Principles of Biomedical Ethics*.

35. Savulescu, "Bioethics," 31–32.

36. Religion in Savulescu's analysis is an ersatz philosophy. Far from Wittgenstein's appreciation for the honest clarity of mysticism within traditional religions, religion and faith for Savulescu are pejoratives.

37. Savulescu, "Clone Me Up, Scotty."

38. Engelhardt, *Foundations of Bioethics*, 56.

This gloss of Savulescu sets up an interesting scenario: many in bioethics seek to marginalize Savulescu for the wide array of his unpopular positions, such as challenging the therapy-versus-enhancement distinction, welcoming PEDs in sports, advocating for pharmaceutical moral enhancements, and openly embracing eugenics. Savulescu comes to these positions by staunchly denying the "natural" principlist values that undergird mainstream bioethics, and within the possible theorizing approaches to bioethics Savulescu's utilitarian bioethics is surely as legitimate as principlism. Both sides claim to be rationally self-evident. For Savulescu, working within the lineage of his mentor, Peter Singer, good philosophy is utilitarian philosophy. In short, while the foundational commitments of Savulescu's utilitarianism are hyperprovocative to many, leading to numerous reprehensible moral postures, Savulescu remains helpful in pointing out the errors of secular, scientific theorizing that assumes natural principles. Principlist theorizing in this mould exists within the shallow language game of bioethics birthed in the second half of the twentieth century with seemingly little awareness of the need to consider its own philosophical and biopolitical commitments.

The notable irony, of course, is that Savulescu's utilitarianism also exists within a system of philosophical and biopolitical commitments that he neglects to acknowledge. Savulescu does not begin with mid-level principles but rather with an assumption of consensus or agreement on how to rank or compare consequences. Following Engelhardt, consequentialist and utilitarian accounts of rationality presuppose an antecedent and even an "authoritarian means of judging, ranking, or comparing benefits and harms." That Savulescu can point to a greater benefit is not decisive given that the issue centers not on "facts" but on the judgments and evaluations of those "facts." What must be stressed is that "science is not an ahistorical deliverance of the gods. Scientific truth, as all human truth, is historically and culturally shaped."[39] In other words, recalling Wittgenstein's admonition that our "craving for generality" stems from our "preoccupation with the method of science," even while Savulescu criticizes bioethics for ignoring the fact-value distinction, in reality he assumes a factual utilitarian judgment that functions as a pseudometaphysical starting point. Hence Wittgenstein reminds us:

> Philosophers constantly see the method of science before their eyes, and are irresistibly tempted to ask and answer questions in the way science does. This tendency is the real source of metaphysics, and leads the philosopher into complete darkness. (*BB*, 18)

39. Engelhardt, *Foundations of Bioethics*, 39; see also 37–65.

For all of Savulescu's critiquing of bioethics's "scientifying" and slavish moralism, he falls prey to his own scientific confusion, and this enacts something of a methodological impasse: Savulescu's good philosophy is utilitarian while the principlism he confronts is purportedly faith-based moralism. Yet is Savulescu's challenge anything more than one additional theoretical approach forcefully proposed amidst the competing theoretical spheres of postmodernity? We must ask why Savulescu's justification of a secular, content-full ethics ought to outrank other approaches. Why utilitarianism in place of intuitionist accounts, casuistic accounts, hypothetical-choice theoretical accounts, rational choice and discourse theoretic accounts, game-theoretical accounts, natural law accounts, or mid-level principlist accounts? Each account "presupposes exactly what it seeks to justify: a particular moral content."[40] Within the amorphous postmodern world, Savulescu's account inherently assumes a rational-moral content, and such moral content will be secured either with "the price of particularity" or universality.

> To have moral content, one must endorse particular moral premises or rules of moral evidence as a point of departure, thus endorsing one from among the class of available moralities. On the other hand, universality is approached at the price of content. For a morality to have content, it must be particular. But which particularity, which moral content, should be endorsed on what basis? Again, to answer such questions, one either begs the question or submits to infinite regress. In the face of these difficulties, no particular moral vision or bioethics, in general secular terms, can be shown to be better than any other.[41]

Savulescu's judgments and valuations of greatest benefits and harms rely upon some greater moral vision that he fails to identify. To grasp more of Savulescu's assumed greater moral vision, it is helpful to see how these waves of scientific theorizing lead particularly into the transhumanist movement. To be sure, transhumanism looks forward as part sci-fi and part technooptimism, and yet the presumed faith in discursive rationality via Enlightenment ambitions remains a steady undercurrent.

40. Engelhardt, *Foundations of Bioethics*, 42.
41. Engelhardt, *Foundations of Bioethics*, 59.

Transhumanism's Enlightenment Roots

> *We live in a society exquisitely dependent on science and technology, in which hardly anyone knows anything about science and technology. . . . Science is much more than a body of knowledge. It is a way of thinking.*[42]

Enhancement and transhumanism are distinct, yet are similar enough in philosophical grounding and assumptions, such that we may discern a common moral vision common to both.[43] Enhancements may be minor and relatively nonintrusive (e.g., caffeine), and the term can be remarkably "elusive" and "conceptually ambiguous," as Michael Bess has argued.[44] While not ignoring these challenges, some scientific and technological enhancements—pharmaceuticals, deep brain stimulators, genetic therapy, and other biotechnological interfacings—are clearly bioenhancements and a means to transhumanism. Hence any strong separation of enhancement and transhumanism is unnecessary, and it is instructive to frame transhumanism as something of an end goal for radical enhancement.[45] That said, Persson and

42. Sagan, "Why We Need to Understand Science," 265.

43. Savulescu has advocated specifically for "moral transhumanism" (Persson and Savulescu, "Moral Transhumanism") and yet does not call himself a transhumanist. "I have some sympathy for [transhumanists] and I think it's great that they're out there pushing that line of argument. I'm not a transhumanist or a post-humanist. I think it starts to take on characteristics of a religion and is a kind of belief in itself. But the ideas are interesting and need to be taken seriously. I wouldn't put all my eggs in their basket, but I'd put some eggs in their basket. The capacity for technology to increase in power is exponential; the capacity of humans to control it doesn't increase exponentially. We have to realize that the technology we've created has reached a point of being runaway" (as quoted in Paulson, "Philosopher Who Says," para. 21). While Savulescu may not consider himself formally a transhumanist, I contend that he has enough "eggs in their basket" to align broadly with the philosophical assumptions in transhumanism that seek through radical enhancements to transcend the limits of the human body and brain. Likewise, we may notice the overlap between Savulescu's understanding of moral enhancement and Fukuyama's assertion that the fundamental tenet of transhumanism is "that we will someday use biotechnology to make ourselves stronger, smarter, less prone to violence and longer lived" (Fukuyama, "'Transhumanism,'" para. 3). In short, I propose to consider radical enhancement and transhumanism carefully to avoid false caricatures but rather to uncover the philosophical stance that I contend undergirds Savulescu's enhancement and transhumanism.

44. Bess lists six types of conceptual ambiguity with the term enhancement: "normal or species-typical functioning, therapeutics or healing, natural functioning, human nature, authenticity and the ambiguity between 'more' and 'better'" (Bess, "'Enhanced Humans' vs Normal People," 641, 641–55).

45. Moreover, if one agrees with Fukuyama's well-known assertion in 2004 that transhumanism was one of the world's most dangerous ideas, the stakes are raised for both transhumanism and radical enhancement.

Savulescu have also invoked transhuman and posthuman technologies as a potential means toward moral enhancement, reflecting their prioritization of "moral humanity" over "biological humanity." For humans to leave behind their human biology is a worthy transition if it effects Persson and Savulescu's vision for a moral humanity.[46]

Following James Hughes—co-founder with Nick Bostrom of the Institute for Ethics & Emerging Technologies—transhumanism espouses the belief that science and technology can be used to enhance and "transcend" the limits of the human body and brain. According to Bostrom, "transhumanism has roots in secular humanist thinking,"[47] and he defines transhumanism as "the intellectual and cultural movement that affirms the possibility and desirability of improving the human condition through applied reason, especially by developing and making widely available technologies to eliminate aging and to greatly enhance human intellectual, physical, and psychological capacities."[48] Describing transhumanism's appeal to reason, Hughes notes:

> One of the main transhumanist blogs is Less Wrong, started at Oxford University under the aegis of transhumanist philosopher Nick Bostrom, dedicated to "the art of refining human rationality." One of the frequent contributors there is Eliezer Yudkowsky, an autodidact writer on artificial intelligence and human cognitive biases who also is the founder of the Singularity Institute for Artificial Intelligence. Yudkowsky has said that one of his goals is to lead a "mass movement to train people to be black-belt rationalists."[49]

Addressing the question "What do we mean by 'rationality?,'" Yudkowsky speaks of the two forms of rationality:

> 1. Epistemic rationality: systematically improving the accuracy of your beliefs. 2. Instrumental rationality: systematically achieving your values. The first concept is simple enough. When you open your eyes and look at the room around you, you'll locate your laptop in relation to the table, and you'll locate a bookcase in relation to the wall. . . . Epistemic rationality is about

46. Persson and Savulescu, "Moral Transhumanism," 668.

47. Bostrom, "Human Genetic Enhancements," 493.

48. Bostrom, "Transhumanist FAQ," #2. Bostrom further defines transhumanism as "the study of the ramifications, promises, and potential dangers of technologies that will enable us to overcome fundamental human limitations, and the related study of ethical matters involved in developing and using such technologies."

49. Hughes, "Contradictions," 625.

> building accurate maps instead. This correspondence between belief and reality is commonly called "truth," and I'm happy to call it that. Instrumental rationality, on the other hand, is about steering reality—sending the future where you want it to go. It's the art of choosing actions that lead to outcomes ranked higher in your preferences. I sometimes call this "winning."[50]

This hope in the rational powers of science aligns transhumanism well within the framework of Enlightenment optimism for human progress and belief in the supreme power of reason. This link, however, Hughes admits, also implies that transhumanism stands to inherit the "internal contradictions and tensions" of the Enlightenment tradition.[51] Simply put, reason is not self-legitimating. The project of reason *qua* reason rules out all foundational premises, including the premise that reason is superior to unreason. "Consequently transhumanists, like Enlightenment advocates in general, need to defend our values with *nonrational a prioris*. Unfortunately, some transhumanists continue to advocate a naïve conception of pure rationality as an end in itself."[52] Any attempt to clear all nonrational a prioris out of the way inevitably leads to a "philosophical dead end" and leaves one within a circle of self-referential appeals to a methodology (rationalism) when facing foundational, ontological questions.

Further, such a hope in the supremacy of reason fails to account for the numerous challenges to rationalism that followed on the heels of the Enlightenment or from within the Enlightenment era with Rousseau's prioritization of primitive nature. Early nineteenth-century Romanticism, the Counter-Enlightenment movement, and others contested the supremacy of rationalism, and of course various strands of postrationalist thinking continued on into the nineteenth and twentieth centuries, notably including Nietzsche, Kierkegaard, Wittgenstein, Foucault, and Derrida, all of whom in their respective ways "represent the implosion of Enlightenment reason."[53]

Beyond scientific rationalism, this alignment of transhumanism with the Enlightenment project may be seen on a number of levels. The primary point for now is Hughes's contention that transhumanism is a direct "product" of the Enlightenment, hence reflecting a "myriad of tensions and contradictions" primarily inherent within Enlightenment rationality. But as Fabrice Jotterand notes, the continuum between Enlightenment

50. Yudkowsky, "What Do We Mean by 'Rationality?,'" para. 1.

51. Hughes, "Contradictions," 622.

52. See Hughes, "Contradictions from the Enlightenments Roots," 622 (emphasis original).

53. Hughes, "Contradictions," 624.

thought and transhumanism needs clarification for there is much more than an extreme optimism in rationality at stake in transhumanism. Transhumanism has "reformulated" key tenets rejected by Enlightenment philosophers, including questions on:

1. Theology: scientific or materialist theology replaces traditional theology;
2. State power: technocratic authoritarianism operates as a new form of state power;
3. Teleology: self-directed evolution toward a technocratic future replaces teleology;
4. Moral authority: the UN Declaration of Human Rights replaces traditional sources of moral authority, that is church authority;
5. Immortality of the soul: the hope to be able to upload the brain on a computational substrate represents a new conceptualization of the religious notion of the "immortality of the soul." [Hence] transhumanism is not only an heir of the Enlightenment but also a radical move away from one of its core values: the autonomy of reason.[54]

In short, while Hughes firmly contends that transhumanism is a product of the Enlightenment, Jotterand points out that the "tensions" and "ideological variants" that Hughes describes in transhumanism qualify how we ought to describe the Enlightenment rationalism of transhumanism.[55] Transhumanism cannot be said to adhere fully to the core Enlightenment value of the *autonomy* of reason but rather holds to an *essential* rationality, and, as the five qualifications above make clear, this rationality is embedded within a core set of philosophical assumptions and practices. Foundational values, presuppositions, cultural practices and historical context always surround and intermingle with "pure rationality."

Interestingly, as Hughes emphasizes transhumanism's attempts at pure rationality, a sermon he delivered to the Unitarian-Universalist Society in 2012 offers an example of the inevitability that scientific reason exists within some greater life context or *Lebensform*.[56] Hughes speculates on the notion of "morality in a pill," wondering how clearly our scientific rationality might guide our ethics. Taking up neuroscience as a concrete, modern appropriation for "age-old religious questions about what it means to be a good person," Hughes asks if the whole idea of virtue, vice, and sin

54. Jotterand, "At the Roots of Transhumanism," 618.

55. Jotterand, "At the Roots of Transhumanism," 618.

56. Hughes, "Morality in a Pill?"

is now anything but "old-fashioned and intolerant," or if the project of optimistic moral enhancement using neuroscience simply hides the essential question.[57] Following MacIntyre seemingly implicitly, Hughes ponders if such moral neuroenhancement can avoid the question: Whose version of morality will we be enhancing? Hughes takes up Anders Brevik as an example. The Norwegian mass killer Brevik insisted that his use of terror to "awaken Norway to the dangers of Muslim immigration" was sane and moral. For late liberals, the question is how to evaluate competing language and rival definitions from others who also claim a reasonable morality. Hughes identifies in liberalism two central moral intuitions—"don't hurt people and don't cheat"—and yet what of rival moral intuitions supporting nepotism, racism, nationalism, and/or the fear of strange immigrants, all of which may be understood as natural within a Darwinian impulse for self-preservation?[58]

In these ways, Hughes questions the universality of moral intuitions or rational moral definitions, but in the end he returns to the liberal safe ground of "some areas" of agreement such as self-control, regulating behavior, etc. This remains inconsistent with his initial challenge, and the closing suggestion of his sermon is that Unitarian Universalism is a "fertile ground for the exploration of ways that these new tools [of neuroenhancing morality] can be used within a liberal model of virtue and the good life, and as a place to stand against the inevitable efforts to use these tools to close down the free and responsible search for truth and morality."[59] In a highly confusing manner, Hughes embraces transhumanism's appropriation of Enlightenment rationality while simultaneously challenging liberalism's attempt to hide its moral biases. Amidst this confusion, however, he raises questions that Savulescu fails to even acknowledge in his equivocation that utilitarianism is good philosophy.

Savulescu's Nietzschean Language Game

> *We can now state Nietzsche's ethic. I think what follows is a fair analysis of it: Victors in war, and their descendants, are usually biologically superior to the vanquished. It is therefore desirable*

57. Hughes, "Morality in a Pill?" para. 1.

58. Hughes, "Morality in a Pill?" para. 5.

59. See Hughes, "Morality in a Pill?," para. 24. It is worth noting the irony in the rhetorical occasion of a *sermon* for addressing the possibilities of stand-alone scientific rationalism. That said, the irony is largely lost within liberal religions that speak in theological language to support whatever the culturally relevant values of the age are.

> *that they should hold all the power, and should manage affairs exclusively in their own interests.* —Bertrand Russell[60]

Savulescu begs the question he should answer—the question of foundational moral content—by repeatedly positing his philosophical method *as* good philosophy. In order to select the best possible outcome for benefits over harms, Savulescu must already know which outcomes are more beneficial and which are more harmful. In other words, "consequentialist ethics presupposes a nonconsequentialist ethics,"[61] and here we can see how the later Wittgenstein's language games offer a clarifying therapy for Savulescu:

> We want to say that there can't be any vagueness in logic. The idea now absorbs us that the ideal "*must*" occur in reality. At the same time, one doesn't as yet see *how* it occurs there, and doesn't understand the nature of this "must." We think the ideal must be in reality; for we think we already see it there. (*PI*, 101, emphasis original)

> One predicates of the thing what lies in the mode of representation. We take the possibility of comparison, which impresses us, as the perception of a highly general state of affairs. (*PI*, 104)

In *PI*, 101, Wittgenstein is pointing to an ideal order for language, a standard or norm that we presume "must" exist because of our use of language. This does not necessarily preclude any such standard, but Wittgenstein's first point is to examine why we think such a norm exists, and how that norm may align with language. Savulescu employs a certain logical sequencing and assumes that the moral content will be revealed through his logical methodology. Is his methodology logical? Certainly, but that is not the question; the question remains as to how we ought to rank the scale of values. *PI*, 104 takes us closer to the point by bringing to attention the "general state of affairs." *How* do we determine the general state? Wittgenstein would have us begin by removing the "*preconception* of crystalline purity" in logic and language (*PI*, 108, emphasis original), and following this touchstone of humility in the sphere of human knowledge, we can then seek not new "*explanations*" or "hypothetical" theories but rather "to assemble what we have long been familiar with" (*PI*, 109, emphasis original). Following Wittgenstein's language games, hence may be taken as a gesture to something mystical or noetic, but before we pursue that depth, we must understand more of the hidden preconceptions and presuppositions within Savulescu's language. What are the reference

60. Russell, *History of Western Philosophy*, 695.

61. Engelhardt, *Foundations of Bioethics*, 47.

points upon which his scale of utilitarian values is fixed? Or, to capture another of Wittgenstein's aphorisms, what is the "picture" or image that holds Savulescu captive and confused (*PI*, 115)?

Now we are better able to examine Savulescu's core commitments as postmodern offshoots of transhumanism, stemming from transhumanism's (mostly) Enlightenment roots.[62] Far from neutral utilitarianism, is Savulescu's transhumanism positing (1) a scientific or materialist theology versus traditional theology, (2) a new technocratic authoritarian state power, (3) a self-directed evolutionary teleology, (4) a moral authority, such as the UN Declaration of Human Rights vs. traditional sources of moral authority, and (5) a hope for immortality via uploading the brain on computational substrate versus the immortality of the soul?[63] If so, we have many deeply substantive assumptions to uncover within utilitarianism as simply "good philosophy." Such robust tenets of transhumanism should be taken as nothing less than examples of late modernity morphing into postmodernity. As Charles Taylor describes it, the modern era from the seventeenth century on is marked by the transition into "modern freedom" that breaks loose from "older moral horizons."[64] Similarly, Brent Waters characterizes this "late modern landscape" by first tracing two cultural shifts: the shift into modernity that can be described as a transition from "providence to progress" while the shift into postmodernity can be described as a transition from "progress to process."[65] Doubtless, these are broad strokes to summarize enormous para-

62. Here again, following Wittgenstein's gestures, we can read Savulescu's radical enhancement as a search for uncovering true nature and humanity. A common argument is that enhancement does not take away from humanity but rather enhances humanity. All well and fine, but such an argument is further evidence of a self-referential language game. Such an argument traces the "frame" of the picture while ignoring the question of "nature" itself.

Tractatus Logico-Philosophicus (4.5): "The general form of propositions is: This is how things are." That is the kind of proposition one repeats to oneself countless times. One thinks that one is tracing nature over and over again, and one is merely tracing round the frame through which we look at it (*PI*, 114).

63. Jotterand, "At the Roots of Transhumanism," 618.

64. Taylor, *Ethics of Authenticity*, 3.

65. Waters, *From Human to Posthuman*, 11–18. It is helpful to specify further Waters's descriptions of modernity as progress-oriented and postmodernity as process-oriented. Engelhardt's explication of modernity identifies two senses of the term "modernity." First, in the period after the Middle Ages—specifically after the Renaissance and Reformation—Western European culture asserted its "powers over the ancient world with a self-confidence born of its successes in exploration, astronomy, anatomy, physics, chemistry, and art." Modernity, in this first sense, is identified chiefly with the Enlightenment as well as Immanuel Kant's proclaimed motto of the Enlightenment: *Sapere aude*! "Have courage to use your own reason!" The second sense of the term modernity connects with the term postmodernity. Particularly from Nietzsche

digm shifts and changing discourses of theology, science, and technology, but by the mid-nineteenth century we can hone in on a particular cultural sense that the Enlightenment's "most prized legacy" of *progress* had come to an end. As Waters points out, the marriage of Baconian utopian thought with a refined Cartesian methodology spurred "rapid industrialization" that fundamentally changed cultural expectations for applied science; beyond progress, a new expectation led to ontologically charged notions of processing nature into something beyond the natural.[66] Equally, we could look to the "corrosive effects" of the "pliable and adaptive" Darwinian natural world as well as the new complexity and instability stemming from Freudian psychology.[67] The external and internal worlds of late modernity were shifting, imparting momentum to a haunting specter:

> Evolution was an account of continuous and violent competition. . . . The drama of evolutionary ascent was predicated on a subplot of death, and the omnipresent threats of pain and misery accompanying it. Moreover, humans were not spared a role in this bloody spectacle, but were shaped by it. As psychoanalysis revealed, humans carried within their subconscious a host of virulent instincts and murderous impulses that the rational mind was hard pressed to keep under control.[68]

Amidst these morphing discourses, morality no longer remains as a noble goal for human rationality but instead functions as one additional adaptive mechanism for survival. Enlightenment optimism in progress shatters before the human will and growing technological power to direct human evolution into new frontiers. As Taylor describes it, these were disenchanted frontiers staked out through "instrumental reason" that prioritized individual lives apart from any broader vision, leaving Nietzsche's pitiable "last men" as the comfort-seeking nadir of modernity's decline. On the political level, Taylor points to de Tocqueville's assessment of "soft" despotism where individuals are "enclosed in their own hearts" and prefer to stay at home rather than actively participate in self-government. Filling this vacuum, "institutions and structures of industrial-technological society" arise that severely influence the course of society, limiting liberties

---

forward, modernity indicates a "strongly adversarial culture" that was primarily anti-Christian and brings "reason itself into question." "With postmodernity the West becomes posttraditional—it abandons central elements of some fifteen hundred years of its culture" (Engelhardt, *Foundations of Bioethics*, 23).

66. Waters, *From Human to Posthuman*, 12.

67. Waters, *From Human to Posthuman*, 15.

68. Waters, *From Human to Posthuman*, 16.

and enabling moral and environmental disasters such as the ones that Savulescu includes in the alarmist rhetoric of his global technodestruction narrative. But where Savulescu calls for processes of moral enhancement, Taylor harkens to a premodern "ethical authenticity" that sees human beings as necessarily united to "God" or to the idea of the good as the essential way to full being.[69]

Connecting these themes of late modernity into postmodernity, we can see how Nietzsche aligns with Savulescu's transhumanist vision, presenting an implicit but critical reference point within his language game. Savulescu is typically identified as a stringent utilitarian, but on a more fundamental level he operates within a postmodern world that fully accepts processes of willful becoming:

> As Darwinism makes clear, there is no fixed human condition given the radical nature of change over time. The fundamental fact of human life is its historicity. Yet philosophy has tried perennially to demonstrate some permanent aspect of human nature. According to [George] Grant, this is precisely the point Nietzsche rejected, insisting that any sense of permanence must be expunged from any vantage point claiming to be modern. . . . Consequently we are forced to accept the "finality of becoming."[70]

Waters proceeds to note Nietzsche's insistence that studying nature yields no moral purpose or "progressive trajectory" for there is no permanence to be discerned through reason. To this, we can align Savulescu's desire to move beyond what he calls the traditional modes of enhancement such as education, opting to change nature rather than study it. Nietzsche's historical sense undercuts traditional morality, as the only meaningful life is one that confronts and overcomes the "historical horizons," horizons representative not of "providential or progressive order" but of values constructed by human will.[71] God is dead, and so, too, objective horizons are lost before historical horizons. "Untruth" is a condition of life for Nietzsche as "the falseness of a judgment" is no inherent objection to that judgment. What is supreme over truth or falseness is the "life-promoting, life-preserving,

69. Taylor, *Ethics of Authenticity*, 26.

70. Waters, *From Human to Posthuman*, 24.

71. Waters, *From Human to Posthuman*, 25. On Nietzsche's historicism, Waters states: "There are two general meanings of the term history. One is a study of the past, while the other connotes a form of existence. In the latter instance, humans are historical beings in that they are shaped by the temporal processes they construct. To be historical in this sense means that past, present, and future are collapsed into a single mode of being" (23).

species-preserving" awareness that logic and synthetic a priori judgments are fictional realities. Moving beyond such "value feelings" is the move beyond good and evil.[72] Truth stems from the will, and the will produces unspeakable violence as Nietzsche's "last men" seek revenge upon nobility while nihilists lash out against a world "where there is nothing noble to will."[73] The possible way between these two poles lies with the hope of the *Übermensch* who embraces fate (*amor fati*) to move beyond revenge and despair courageously to remake the world.

How realistic this hope is for Nietzsche is a question for another time, but the primary point to draw from Nietzsche's prophetic critique is the postmodern embrace of process and flux that is mastered and perpetually remade by willful humans. This postmodern world enables the transhumanist ambitions for materialist theology, technocratic states, a self-directed evolutionary *telos,* and more. The synthesis of science and technology perpetually furthers this orientation of "radical plasticity," and here we can discern a difference between Nietzsche's hopeful *Übermensch* and transhumanism's self-directed *telos*.

> To will, rather than love, a fabricated fate is accompanied by a recreation of the one who is willing. There is no need to wait for the *Übermensch* to evolve, for a similar being is to be created or engineered. . . . The radical plasticity of the postmodern orientation is attained because its *telos* is also its *techne*. The future will largely be what we make of it; what it is willed to be. The postmodern *telos* of no *telos* is in fact a *telos* of *techne*.[74]

Moreover, this *telos* of *techne* is ontologically charged as the postmodern world creates and is shaped by a cybernetic reality. As our technological dependence grows, the distinctions between "natural" and "technological" dissipate, and humans morph into cybernetic organisms. Stephen Garner suggests that under the influence of a technological society, one need not have technology physically embedded within one's body to be considered a cybernetic organism. Rather, the boundaries between organism and machine are fluid within today's social spheres. Hence the term "cyborg" operates on a biological and cultural level, highlighting that "technology can no longer be thought of as simply tools to be applied, but rather is the environment in which we live and breathe and have our being."[75] As Wa-

72. Nietzsche, *Beyond Good and Evil,* 201–2.

73. Waters, *From Human to Posthuman,* 26.

74. Waters, *From Human to Posthuman,* 30.

75. "The biological cyborg, in which human beings are literally fused with their technology, is a powerful and often violent portrayal of human beings in their

ters notes, "to cross the postmodern divide is also to enter the posthuman terrain," and this characterization can be further described under three attributes or headings: subjectivity, malleability, and mastery. The postmodern subject, existing within a cybernetic order, is marked by the primacy of mind such that within this mentalist prioritization, postmoderns are "embodied minds." Personhood and autonomy within the late liberal tradition exist on a spectrum of rationality, and this modernist spectrum is expanded further in transhumanism. Within a cybernetic reality, the boundaries of time and space are compressed, and self-directed enhancements such as psychopharmacology, prosthetics, and even neural implants are options for upgrading the body, as well as uploading and updating the mind. The net result is that "the person is defined and expressed through self-referential experience that transcends the body in order to maximize the range and quality of one's potential subjectivity."[76] To understand the posthuman attributes of malleability and self-mastery, we need only further trace the implications of prioritizing information and virtual reality over materiality. If information and data trump materiality, the body is perpetually a hindrance and a poor substitute for other superior computer mechanisms that offer a "superior substratum." Biological embodiment is not essential but accidental as "the autonomous person is not a given, but a socially constructed individual with malleable, rather than definitive borders."[77] Reshaping malleable persons will always take place within some particular set of goals. Hence nature is a raw resource (*Bestand*) awaiting mastery by the human will.

What is clear in Savulescu's arguments for radical enhancement and the above descriptions of transhumanism is that both share a common

relationship with their own technological proclivity, and one that generates immediate response. The cultural representation of the cyborg seems more benign because it does not shock through forcing a confrontation with that human agency. However, this less common cultural use of the term highlights the extremely complex network of relationships between human beings and the products of human agency, which is sometimes obscured by the core common imagery of the biological cyborg. The image of the cyborg, both biological and cultural, and the hybridity it represents, is an ambivalent one in the world of shifting boundaries, where technological manipulation of the world relies upon there being some kind of consistent structure or order to affect, and yet in doing so existing notions of some kind of fixed ordering of the natural world are challenged or rejected" (Garner, "Hopeful Cyborg," 89–90).

76. Waters, *From Human to Posthuman*, 37; 33–40. Responding to Habermas's late liberal society, Waters points out that "embodied personhood is no longer a defining boundary, but an unfortunate limitation on the scope and quality of subjectivity; a limitation which should be overcome, not honored . . . What Habermas fails to recognize is that late liberal society is no longer comprised of embodied persons but populated with persons who happen to have bodies" (39).

77. Waters, *From Human to Posthuman*, 43.

moral vision. Savulescu can be confusing as he calls for "citizen norms" to increase the "prosperity of a world community," but then, under the conviction that such efforts are too little too late, he places his hope in any possible pharmaceutical, biomedical, or other moral enhancement. Savulescu critiques principlist, "scientified" bioethics, but then employs his own scientific rationalism. He prioritizes "moral humanity" over "biological humanity," embracing transhumanism as a likely way to achieve greater benefit.[78] In these ways, Savulescu exhibits a mixture of modern and postmodern elements as he vacillates between whatever means are available. "Postmodernity is simultaneously the affirmation and negation of modernity,"[79] and Savulescu seems precisely at home in such a mangle.

Navigating these enhancement and transhumanist moral visions is challenging, for it is not always clear exactly what conceptual claims are being made. What is the specific teleology of process-driven, self-directed transhumanism? Who defines the harms and benefits available through evolutionary moral transhumanism? The central irony is that amidst these ambiguities, Savulescu is the one leveling claims of "conceptual confusion" against his colleagues in bioethics who fall prey to principlism as a "lethal and widespread malaise."[80] Within the principlism game of bioethics, Savulescu is helpful for critiquing the errors of scientific theorizing, but he remains ensconced in a self-referential logic. He has challenged principlist bioethics and raised the stakes, but in the end his challenge offers little more than a new language game.

Moral Enhancement Serving the Posthuman God

> *It was [Bacon's] The New Atlantis that first canvassed the possibility of a joint series of operations that would combine a new system of scientific investigation with a new technology. At a moment when the bitter struggle within Christianity between contentious doctrines and sects had come to a stalemate, the machine itself seemed to offer an alternative way of reaching Heaven. The promise of material abundance on earth, through exploration, organized conquest, and invention, offered a common objective to all classes.* —Lewis Mumford[81]

78. Persson and Savulescu, "Moral Transhumanism," 667–68.
79. Waters, *From Human to Posthuman*, 49.
80. Savulescu, "Bioethics," 28.
81. Mumford, *Myth of the Machine*, 283.

Drawing together these themes, what undergirds this new language game is a transhumanist vision possessing a strong Nietzschean "sense of dismay" that humans may not grow into something greater, willing themselves beyond the historical horizons. At the same time, as Gerald McKenny notes, transhumanists depart from Nietzsche when they acknowledge that technology enables a human future beyond what can be conceived of as human. Hence McKenny imagines Nietzsche asking: "what could motivate us . . . to pursue such a good, except some debilitating form of human self-hatred?"[82] In this way, the *Übermensch* is a helpful comparison to the ideal transhuman, but where the *Übermensch* accepts fate as furthering a more powerful humanity, the transhuman vision wills beyond bodily boundaries in an effort to realize the next step in human evolution from ape to human to posthuman.[83] In place of *homo sapiens* formed within traditional boundaries of flourishing, the goal is to transition beyond boundaries into some sort of *techno sapien* identity. These posthumans would cease to resemble humanity in any meaningful way as they imagine "completely synthetic artificial intelligences" or other "profound augmentations of a biological human." In short, transhumanists

> yearn to reach intellectual heights as far above any current human genius as humans are above other primates; to be resistant to disease and impervious to aging; to have unlimited youth and vigor; to exercise control over their own desires, moods, and mental states; to be able to avoid feeling tired, hateful, or irritated about petty things; to have an increased capacity for pleasure, love, artistic appreciation, and serenity; to experience novel states of consciousness that current human brains cannot access.[84]

82. McKenny, "Transcendence, Technological Enhancement, and Christian Theology," 187.

83. Hauskeller has suggested further qualifications for aligning Nietzsche with the posthuman, leading to his conclusion that Nietzsche is not "a major inspiration for transhumanism." Hauskeller points to (1) Nietzsche's disdain for "anyone seeking to improve the human condition"; (2) transhumanism's desire to reevaluate "certain"—and not all—aspects of our existence; (3) contra Nietzsche, transhumanism continues to see our humanity seated primarily in the mind; and (4) transhumanism seeks immortality while Nietzsche decries such optimism. These qualifications for a close connection aside, Hauskeller admits there may be some common ground, and for the purposes of this chapter the association described by McKenny ("Transcendence, Technological Enhancement, and Christian Theology"), Bishop ("Transhumanism, Metaphysics, and the Posthuman God"), and Waters (*From Human to Posthuman*) is more than reasonable (Hauskeller, "Nietzsche, the Overhuman and the Posthuman," 5–8).

84. Bostrom, "Transhumanist FAQ," #5–6.

Admittedly, these claims can seem farcical, and, as Bishop has noted, a "pitfall" danger is to take these seemingly "Polly-Anna-ish" claims too seriously. Transhumanists such as Bostrom and Hughes, however, are far from insincere as they take seriously "the spirit of Bacon's statement that the purpose of knowledge (and its technological fruits) is to relieve the human estate . . . to transcend human frailties, not by relieving the human condition of its frailties, but by relieving us of the human condition itself."[85]

Within this transition to the posthuman, what most notably emerges is an ordering theology, a theology that Bishop explicates in light of Heidegger's ontotheology.[86] Ontology is the search for *ousia*, for *whatness*, while theology searches for a more particular existence, for *thatness*, and for Heidegger the passing of each historical epoch is best understood in terms of the interplay between asking (and confusing) questions of ontology and theology.

> [E]ven Nietzsche, who is much less concerned with being and more concerned with becoming, thinks theologically when he "thinks the *existentia* of the totality" by proclaiming the "eternal return of the same"; after all, eternal recurrence is not just "the way that the totality of entities exists . . . but also their *highest* mode of existence (as the closest the endless stream of becoming comes to *being*)."
>
> Thus, for Heidegger the history of metaphysics is a history of founding ontotheologies, which were unable to secure their own ground. The history of metaphysics, then, is a history of swinging between foundation and abyss, with the overturning of a previous ontotheology by the next ontotheology. In other words, for a time ontotheologies give "a perhaps necessary appearance of ground."[87]

Nietzsche's Dionysius, the god of chaos, represents the struggle for power, and yet while Nietzsche thinks theologically in some ways, the will to power is not "conscious agency" aimed at a certain end but rather exists

85. Bishop, "Transhumanism, Metaphysics, and the Posthuman God," 701.

86. Heidegger's analysis of ontotheology is well known. Ontotheology names "two stalks" of metaphysics as ontology—that which names the *ousia*, the protosubstance, that takes on various historical modes—and theology—that which asks for some particular understanding of *ousia* as "*which*" entity or in "what sense" it is (Bishop, "Transhumanism, Metaphysics, and the Posthuman God," 704, emphasis original).

87. Bishop, "Transhumanism, Metaphysics, and the Posthuman God," 704 (emphasis original). Bishop is quoting and interacting with Thompson, *Heidegger on Ontotheology*.

as a "nonmentalist or nonconscious becoming—will without agency."[88] Even so, we inevitably seek out rational reasons for why we are the way we are, which amounts to a confusion of "attributing final causes" where only the "being of entities" exists for Nietzsche. This illustrates a point where transhumanism seeks to depart from the "will without agency" analysis and to overstep the "natural" processes of evolutionary selection as rational beings, beings who seek to discern and willfully remake nature into the ever-becoming. Here the *telos* of *techne* synthesizes with the human will, and in turn "the human will, an evolutionary achievement, turns to order the chaos of creative ontology, and thereby enacts an ordering theology." In short, the ordering force of the human will operating upon the powers of creation is transhumanism's theology:

> [T]he god of these transhumanist philosophers is the god that orders the creative power toward a new being, a new god, that is to say toward the posthuman. Transhumanist philosophies, in my opinion, are the coincidence of eternal and creative forces of becoming just as they turn in the conscious moment toward control, toward mastery. Transhumanism seeks to differently embody the *Übermensch.*[89]

This theological point is critical. Building upon an analysis of language games in theoretical bioethics, we can now glimpse within radical enhancement and transhumanism a theological impulse that is far from Savulescu's naïve calls for "good" secular philosophy. This is a postsecular landscape where theological thinking is revealed within all ethics that must admit to some moral vision that grounds the system of language employed within various strands of bioethics.

As Heidegger's ontotheological critique shows, a confusion exists amidst unidentified strands of philosophical and theological thinking. Moreover, following George Grant and Simone Weil, Waters comes to this same point in a different way. Proponents of a posthuman future assert a "religious claim on philosophical and ideological grounds" by confusing the difference between necessity and goodness. Necessity represents normative requirements for human life such as air, water, food, movement, shelter, reproduction, etc., and these necessary elements are "formative" in our cultural (*cultus*—systems of worship) habits. As this last sentence evidences, the tension between the ontological and theological already arises. Moreover, as Waters argues, necessity presents two problems in that our necessary elements (1) reveal creaturely finitude and temporality, and (2)

88. Bishop, "Transhumanism, Metaphysics, and the Posthuman God," 705.

89. Bishop, "Transhumanism, Metaphysics, and the Posthuman God," 707, 705–7.

these necessary elements that are acquired, used, and consumed are equally finite and temporal.[90] Meeting the essential necessities for human existence and cultivating habits amidst and beyond the barest necessities are features we recognize in *homo sapiens*, and yet beyond necessity humans are moved by visions of the good. Hegel and Marx stand as examples of philosophical strategists who synthesize the necessary and the good as historical phases and economic classes move forward toward realizing some greater good through dialectical progress. Attempts to transform necessity into goodness frequently lead to a "cavalier attitude toward evil, for acts of cruelty and violence are justified by historical necessity."[91]

Within this distinction of the necessary and the good we can see radical enhancement and transhumanism standing in opposition to, even denial of, necessity. Limitations and inherent boundaries are aspects of the main problem for transhumanism, and if the solution lies in abolishing the formerly held understanding of finite human nature, we may easily see where immense violence and abuse will be justified under the aegis of posthuman progress. With this assumed posthuman goodness emerging from necessity, an ambiguous *telos* of "some" future good, once established, will almost certainly judge any opposition as antiquated, sentimental conservativism.[92] Indeed, this last claim is a full broadside against those such as Leon Kass who wish to conserve a sense of human dignity based on intuition of what is natural. Such appeals to dignity may take root within a time and place or community where some consensus of flourishing nature can be found within the term. Without some more shared moral vision, however, as Savulescu's critique of bioethics has already revealed, asserting human dignity means little.

90. Waters, *From Human to Posthuman*, 70, 69–79.

91. Waters, *From Human to Posthuman*, 71.

92. A fundamental ambiguity lies beneath the surface of transhumanist visions. What is the nature of the end goal? Ray Kurzweil envisions a future Singularity by 2045 where technology will irreversibly transform human life into a "single mind" that will be a billion times more powerful than human intelligence today. This Singularity will simultaneously remain human yet be nonbiological. Audrey DeGray is a computer scientist who seeks to engineer a therapeutic, antiaging "negligible senescence," a being who does not age. Nick Bostrom focuses on philosophical arguments for radical enhancement. His vision of posthumanism is flexible, and his main opponents are natural humanists such as Francis Fukuyama, Leon Kass, and Bill McKibben, who see humanity and nature as things to be preserved within their current boundaries. James Hughes focuses on social transformations that we may expect from enhancement. See Agar, *Humanity's End*. Amidst these different strategies of transhumanism, how are we to discern the difference between "varying prognostications that run the gamut from utopian dream to apocalyptic nightmare" (Waters, *From Human to Posthuman*, 69)?

What is meaningful is the authentic and explicit admission of theological thinking where a vision of the good is particular and where attempts to name "*thatness*" within *ousia* are not hidden within secular philosophy. Here advocates of a posthuman future fall short, and as they seek to further Enlightenment modes of rationality and human flourishing, they create an "image and likeness" of their own choosing:

> In appealing to a humanistic pedigree, transhumanists betray hidden religious convictions shaping their agenda, for the humanistic tradition they claim as their own cannot bear the heavy, transformative burden they place upon it. . . . Aspiring to become posthuman is tautological, because it seeks to annihilate both the measure and what is measured . . . To assert that humans should become posthuman requires the invocation of a higher or transcendent good that trumps the anthropocentric standard. . . . what is the source of the "trans" that justifies its affixation to "humanist?"[93]

Our fundamental issue up to this point has centered on the confused intersection of science, religion, and theology. Transhumanists start within purportedly neutral scientific reason, yet justify many of their claims on implicitly religious and theological grounds. Further, the Nietzschean moral vision within radical enhancement and transhumanism functions as a "power theology, a subtle theology of the *Übermensch*."[94] A similar, inverse confusion, however, can be seen stemming from explicit theological figures who seek to further theological claims on scientific and technological grounds. Burdett points to Francis Bacon, N. F. Fedorov, and Pierre Teilhard de Chardin as figures within Christian history stemming from the sixteenth to twentieth centuries who affirm technology and enhancement within their Christian visions.[95] Bacon is often held up as a key figure within an overly simplistic narrative that sees technology arising or following from the advances of the scientific method. For some, Bacon's "knowledge is power" is a scientific statement that has been adopted by others in support of unrestrained technocratic utopias, but a closer look reveals a more nuanced, intermingled relationship between technology and science. Bacon's science was technology, and technology and theology both lie within the heart of Bacon's crystallization of the modern scientific method. The *Instauratio Magna* contains six parts—*The Division of Sciences, The Novum Organon, The Phenomenon of the Universe, The*

93. Waters, *From Human to Posthuman*, 78.

94. Bishop, "Transhumanism, Metaphysics, and the Posthuman God," 713.

95. See Burdett, "Transcendence and Human Enhancement."

Ladder of the Intellect, The Forerunners, and *The New Philosophy*—and the title likely alludes to the Vulgate's *instauratio* or to the "restoration" of Solomon's temple and "spiritual edification." The term *instauratio* carries a long, multilayered history, and yet even if some of Bacon's use of the term stems from the Vulgate, we can see a typology or symbolism of instauration in the prophetic rebuilding of Israel as well as an instauration of the cosmos in the apocalypse.[96] For Bacon, both supernatural and ecclesial means are necessary to bridge mankind's alienation from God, and yet the relationship between humanity and creation operates on a different plane. On the natural horizon, human *dominion* is a capacity belonging to Adam, and the *Novum Organon* is the instrument to curb man's temptations toward four "illusions and idols" of fallen human senses, personal prejudices, the marketplace, and false systems of belief. Bacon's vision of instauration sought to employ information to build new technologies that would transform humanity "from a species confined by ancient myth into one that broaches upon the divine."[97] In short, the Baconian method enacts religion, science, and technology as powerful humans form a new "priesthood" with unlimited resources available within the book of creation as raw material. Knowledge is power within the Baconian vision, and yet this power stems from a theological vision of the cosmos.

In a related way, Nicolai Fedorov's ideas may have inspired to some degree the resurrection scene at Ilyusha's grave in *The Brothers Karamazov*, although it is hardly fitting to overalign Dostoyevsky with Fedorov given Fedorov's focus on how humanity may scientifically, even chemically, participate in the resurrection of the dead. Fedorov's primary work, *The Philosophy of the Common Task,* responds to the lack of harmony he saw between people and nature as well as the unbrotherly tension between fellow human beings. He critiqued theoretical philosophers such as Kant for neglecting the needed "action in the world," and he pictured death as the proper enemy of natural and human energies. Many of Fedorov's scientific and cosmological notions pushed the known limits of his age, including weather manipulation to further crop harvests and circumnavigating the globe with immense cables to manipulate magnetic fields. Fedorov's most notable aim, quite bizarrely, was to combat monophysitism by asserting that Christ's *human* nature opens the possibility for humans to participate mechanically and physically in the resurrection. Trinitarian love remains the central impetus for the resurrection, but man's "technological manipulation of matter" presents a synthesizing possibility for Fedorov. Hence "Fedorov's

96. Whitney, "Francis Bacon's *Instauratio*," 371–90.

97. Burdett, "Transcendence and Human Enhancement," 22, 20–25.

entire project hinges upon the utilization of science and technology for the universal and material resurrection of all humanity."[98]

Lastly, Pierre Teilhard de Chardin, is no stranger within writings on transhumanism. Here Burdett points to Teilhard's eschatology that focuses on "social unification," continued future "mechanization"—symbiotically relating technology and consciousness—and a heightened "vision for mankind." Through the processes of "Noogenesis" mankind aims at singularity where the Noosphere is a collective mind of humanity. In a strong gnostic turn, Teilhard's vision of the cosmic Christ entails an abandonment of the material world as the Noosphere develops into full maturity. Naturalistic evolution is deemed a "necessary precondition" for further "divinization" beyond all matter where the "Ultra-human" penultimately appears in glory.[99]

These three explicit theological accounts may seem radically separate from the implicit theological move within transhumanism, a move that directs the power of the will to order the chaos of eternal becoming and process. In reality, however, the space between them is not as distant as one may expect. The accounts from Bacon, Fedorov, and Teilhard seek to exist within some specificity of three distinct strands and eras of specifically Christian theology, and yet it is a farcical stretch to connect their technological visions of enhancement to the concerns of traditional Christian theology[100] that hinges upon the incarnation; put simply, we cannot miss the mystical trajectory that Christ the Son of God became man so that humankind may share in the Divine. Instead, transhumanism autonomously inverts the Christian *ordo* in St. Athanasius's well-known words—"for he was incarnate that we might be made god"[101]—instead to imagine mankind ascending to imagined diety. Whereas the ancient church fathers spoke of *theosis* or deification as the way of mankind's participation and unification with the Triune God, and other authentic religions seek transcendent communion with the gods and beyond human mortality, the path of transhumanism is a self-enacting transition or transformation of the human self where there no longer is a clear picture of what remains as human. Bacon, Fedorov, and Teilhard maneuver their theological ends around their self-authenticating visions of the posthuman in a way that places them well outside traditional Christianity

98. Burdett, "Transcendence and Human Enhancement," 28, 25–28.

99. Burdett, "Transcendence and Human Enhancement," 32, 29–32.

100. Here I would understand "traditional" Christian theology as a prayerful way of life—including the inseparable tapestry of liturgy, mysticism, spirituality, moral dogma, art, and more. This tradition may be seen in the conciliar and ecclesial theology of the first millennium church.

101. Athanasius, On the Incarnation, 107.

and more in the transitional mode of either "self-aggrandizement" or as a cog within biopolitical machinations. As Bishop notes:

> [O]nce we understand beings as the concatenations of forces in our contemporary ontology, one does not transgress, but only transcends human frailty. Yet, the turning of power in the human will to direct these creative evolutionary forces is the theological grounding of our epoch. The greater power of the ordering force must be brought to bear on the chaotic forces. Transitional beings are really directed at some other higher good, whether one's own self-aggrandizement or that of a political power. Bodies, humans are transitory things ordered by a human will with a greater *telos*, and like all transitory things, these things will become materials to achieve that goal. If nothing else, our history has taught us that.[102]

As the transhuman leaves behind humanity, he simultaneously remains caught between the idolatry of self—a self-directed, autonomous tower of Babel for every individual—and the external power forces that shape and remake the self into the image of the *Zeitgeist* of becoming. Bishop furthers the point by noting that Foucault's biopolitics—subtle forces cycling between bodies and politics—removes the real possibility of individuals as such seeking transformation within a free society. The "univocity of Being for our time is that power circulates," which leaves transhumanism afloat amidst historical, social, and political willing, all the while seeking to order the chaos of coercive forces: "power ontology becomes power theology."[103] To this we may add Lewis Mumford's prophecy that the fourth stage of human-technological invention would lead to an evolved technohumanity where the "practical benefits" of machines cannot be accepted without accepting the "moral imperatives and aesthetic forms" of the machine.[104] Whether we trace the effects of external political and social forces, or the new interiority of technologized bodies, what remains evident is that radical enhancement and transhumanism imparts no simple narrative of "good" philosophy or universal principles of bioethics that are effective within secular society. The

102. Bishop, "Transhumanism, Metaphysics, and the Posthuman God," 716–17.

103. Bishop, "Transhumanism, Metaphysics, and the Posthuman God," 717.

104. Mumford, *Techniques and Civilization*, 355. Mumford's first stage is *eotechnics* where water, wood, and other forces of nature were employed; the second stage is the *paleotechnic* phase marked by use of coal and iron in the Industrial Revolution; the third is the *neotechnic* era where electricity and the birth of synthetic compounds came into common usage. Our current biotechnological era, which Mumford calls the *biotechnic* period, is denoted by a "complete integration of the machine with human needs and desires" (353).

notion of discrete, autonomous human beings who may be technologically furthered in their humanity by moral enhancements via pharmaceuticals, neural implants, or some other as yet unknown nanobiotechnological remedy is a phantom and impossibility. What remains is a philosophical and theological interplay amidst these questions that is multivalent and far more complex than Savulescu's call for "good" philosophy to curb the immanent dangers of the environmental collapse or global terrorism.

Rather, what is at stake is a matter of rival *Weltanschauungen* and competing theological modes of thinking that articulate our value scales for biology, technology, and humanity. These questions are essential to bioethics, and it seems increasingly likely that the future of bioethics will continue to embrace enhancement *as* therapy, regenerative medicine *as* the new norm, and the health-satisfaction of Juvin's "new body" *as* the *telos* of medicine rather than the care of the sick and dying. In such a context, understanding the materialistic and scientific theology that centers on a posthuman god is essential for navigating amidst the conceptual confusions and coercive power forces within bioethics. Moreover, this focus on theological thinking necessarily embedded within ethical thinking is a fitting transition to the final section of this book that gestures toward an escape from the fly bottle of immanence and scientific theorizing.

Part III

Escaping the Ethical Fly Bottle

5

Pedagogy in the Ruins: Why Procedural Bioethics Matters

THE GOAL OF THIS chapter is first to reflect upon a notable figure in the field of bioethics, H. T. Engelhardt, who engages bioethics critically yet with a therapeutic aim. This critical-therapeutic strategy is ripe for comparison with Wittgenstein, and aligning Engelhardt and Wittgenstein in this way is the second goal of this chapter. While it would not be fitting to place Engelhardt directly in Wittgenstein's wake, a strong family resemblance arises, particularly in the way that Wittgenstein and Engelhardt both present a critical mapping—of logic/language and bioethics respectively—as a way to argue toward modes of being and thinking that necessarily transcend theoretical and secular reason. In this way, both philosophers call for an ethics beyond the thin possibilities of secular philosophy. However, notwithstanding their general similarities, Engelhardt explicitly moves beyond Wittgenstein's silence in a way that points to a constructive engagement with ethics and bioethics from the particular theological perspective of Orthodox Christianity.

Engelhardt's Narrative of Thin, Secular Bioethics

> *Christianity is not based on a historical truth; rather, it offers us a (historical) narrative and says: now believe! But not, believe this narrative with the belief appropriate to a historical narrative, rather: believe, through thick and thin, which you can do only as the result of life. Here you have a narrative, don't take the same*

> *attitude to it as you take to other historical narratives! Make a quite different place in your life for it.—There's nothing paradoxical about that!* (*CV*, 32)

H. T. Engelhardt is known predominantly as a philosopher of difference, and the differences he points out carry strong implications for the possibilities of a peaceful or orderly existence.[1] One of his primary tactics is to point out the subtle or not-so-subtle violence that can stem from autonomous reason, a rationality that can trick us into "seeing the sameness and the unity in our facile ideas."[2] A number of his works carry references to MacIntyre and Hauerwas, and yet Engelhardt remains less optimistic than either regarding the possibilities of reason and dialogue across different narratives.[3] Bioethics, for Engelhardt, must be envisioned within the family tree of the Enlightenment project that believed in a common rationality able to deliver universal conclusions. Such optimism in theories and common rationality for Engelhardt amounts to an idolatry of reason, and hence his challenge steps broadly in rhythm with established critics of modernity such as Nietzsche, Derrida, Foucault, and Wittgenstein, etc.—all of whom demonstrate that the presumed triumph of reason in the Enlightenment project is vapid, and that the reality behind such idols of reason turns out to be little more than will to power. For Engelhardt, the Enlightenment hope for a secular morality—a way of accounting "in general of what individuals owe each other and ought to do"—stems from a characteristic attempt in Western history "to see reality from the anonymous perspective of reason," and such a normative "view from nowhere and outside of particular history" seeks nothing short of "*the* rational account of being and morality."[4] Faith in reason has replaced faith

1. Though Engelhardt's scholarly focus is bioethics, he sees the way to address bioethics rightly as necessarily running through the deepest philosophical and theological questions. The preface to *The Foundations of Christian Bioethics* opens with these statements: "This book is as much about a philosophical puzzle as it is about bioethics. This book is more about a religious quest than it is about a philosophical puzzle. Yet, it is directed to a philosophical puzzle which it approaches through philosophical reflection and analysis. The philosophical puzzle is this: if we are trapped in immanence, can moral truth be anything but ambiguous?" (xi). Moreover, in addressing these significant philosophical puzzles, Engelhardt does not hesitate to see political or social implications of foundational philosophical or theological commitments. Engelhardt often points to the power struggles and violence in the bloody twentieth century that stems on many points from "real evil," a "hatred and pride" that is beyond mere "intellectual mistakes" in philosophy (Engelhardt, "Foundations of Bioethics and Secular Humanism," 272–81).

2. Bishop, "False Gods," 301.

3. Lucie-Smith, *Narrative Theology and Moral Theology*, 47.

4. Engelhardt, *Foundations of Bioethics*, 4.

in God, leaving bioethics to flounder in the moral ambiguity of life after God.[5] Abstract, universal principles are purportedly deduced apart from any embodied, communal tradition, and such principles are deemed to be self-evident, self-authenticating, exceptionless norms. Here, amidst these modern ruins of the Enlightenment project, the best we can hope for is some degree of honesty that admits that bioethics currently exists primarily on the thin level of the procedural as a *modus vivendi*.

For Engelhardt, the procedural ethic described in *The Foundations of Bioethics* (1996, 2nd ed.) is far from the ideal of a "content-full morality" offering substantive guidance on what is right or wrong; rather, his procedural ethics is founded merely on the consent or permission of rational persons. In a secular society, these rational persons meet as "moral strangers" who engage with one another on the thin moral grounding of consent, whereas "moral friends" are those who share a transcendent authority and guidance for a canonical, content-full morality. Further, Engelhardt differentiates between "communities" of those who are bound by a common moral tradition and/or practice based on a common moral vision of the good life and "societies" of those individuals "who find themselves in diverse moral communities."[6] Embedded within these communities and societies, our encounters with moral strangers and strikingly different moral visions exemplify the "postmodern philosophical predicament" resulting from the failed "modern philosophical project to discover a canonical content-full morality."[7]

Surveying Engelhardt's overarching philosophical narrative and theological critique of secularism—pointing out the disorder and violence in secularism, as well as its implicit, essential theological structure—one can easily see why he is frequently misread and often controversial within bioethics.[8] No one, however, can contest the scope and influence of his scholarship within bioethics given his work alongside figures such as Robert Veatch, LeRoy Walters, Alasdair MacIntyre, Tom Beauchamp, and James Childress in composing the *Belmont Report* and for his early involvement at the Kennedy Institute for Ethics at Georgetown University. Despite Engelhardt's long-standing place within bioethics, his methodology or style has proved a perpetual challenge to many in that he insists on first examining the possibilities for bioethics from within immanent, procedural ethics

5. Engelhardt, *Foundations of Bioethics*, 6; see also Engelhardt, *After God*; Bishop, "False Gods," Cherry, "Emptiness of Postmodern, Post-Christian Bioethics."

6. Engelhardt, *Foundations of Bioethics*, 7.

7. Engelhardt, *Foundations of Bioethics*, 8.

8. See Bishop, "False Gods," and Nash, "H. T. Engelhardt, Jr."

before venturing into any ideal, content-full morality. In short, he argues that thin, immanent, procedural ethics are the current norm, despite claims to thick, common morality and universal rationality. This choice to attend first to the malady before framing the treatment aligns well with Wittgenstein's notion that "the philosopher treats a question like an illness" (*PI*, 255). This point is critical, for numerous readers have mistaken Engelhardt's assessment and critical description of secular ethics as his ideal prescription, supposing that his strategy of stridently carrying postmodern thought to its logical fruition is his proposed philosophical position.[9] To be sure, Engelhardt engages first from within the logic of secular bioethics, but the engagement is intended as a pedagogy of repulsion, revealing how little is accomplished from within the logic of secular consensus:

> [*The Foundations of Bioethics*] does not celebrate the chaos, or even much of the diversity, and surely not the moral perversity and vacuity of this landscape. Instead, this volume offers secular means for coming to terms with the chaos and diversity of postmodernity. The means are meager and offer no transcendent fulfillment. But they are all that is available in general secular terms.[10]

Engelhardt's focus on procedural ethics is no ideal prescription but rather a description of one way—the inevitable way, Engelhardt argues—to navigate bioethics amidst the secular aftermath of the failed Enlightenment project.[11]

Admitting such shortcomings in modern philosophy and noting the limits of procedural ethics as merely "all that is available" is bleak. In his words, "the failure of the modern philosophical project to discover a canonical content-full morality constitutes the fundamental catastrophe of contemporary secular culture and frames the content of contemporary bioethics."[12] Yet amidst this fragmentation, Engelhardt points to the startling

9. Nash, "H. T. Engelhardt, Jr.," 275–77.

10. Engelhardt, *Foundations of Bioethics*, 10. This point is stressed earlier in the preface (xii): "[*The Foundations of Bioethics*] is not a presentation of my concrete moral ideals, my concrete morality, or my concrete bioethics. Quite to the contrary, I regard this book as exploring the possibility for morally authorized collaboration with moral strangers in the ruins of the Enlightenment project. It examines the failure of moral philosophy to deliver a canonical, concrete rationality. It explores the implications of this failure for bioethics. It gives justification for the hope of salvaging something from the wreckage of the project of establishing a secular morality that can bind all. It does not provide a concrete moral vision."

11. Engelhardt, *Foundations of Bioethics*, 13–14.

12. Engelhardt, *Foundations of Bioethics*, 8.

disconnect many bioethicists make between this philosophical disunity and their work in bioethics, proceeding with the task of applied ethics as though "it were obvious which secular ethic ought to be applied."[13] Dialogue on the proper physician-patient relationship, professionalism, justice in health-care distribution, biotechnology, and countless other topics all proceed within the plurality of postmodernity as though all shared a common moral core. Bioethics hence reenacts the Tower of Babel; while no one shares the same language, all carry on building whatever extension is desired to the morphing project of bioethics. In challenging and questioning the grounding for morality in modernity, Engelhardt's argument seeks to convert others away from the trappings of secularism by "exploring the consequences" of secularism and the "irremedial plurality of postmodernity."[14] One can think of this maneuver as being simultaneously pessimistic and realistic, for in starkly describing and exploring secularism, the goal is not to remain within the problematic, postmodern throes of secularism, but rather to turn and flee from such undesirable ethics of procedure.

In short, bioethics functions amidst moral controversies that appear irresolvable in a secular society. The second chapter of *The Foundations of Bioethics* ("The Intellectual Bases of Bioethics") chronicles the struggle to discern amongst various and competing ethics ensconced in ambiguity. Here, for Engelhardt, there is no bio*ethic*; there are only bio*ethics* for we have not "one sense of ethics, but a family of senses."[15] The term "ethics" in modernity is ambiguous as it will vary from community to community (e.g., the Hassidim and the Amish), or in its purpose for rules of conduct amidst professions (e.g., lawyers, accountants, physicians, and nurses, etc.). Furthermore, as a way of organizing various approaches to settling moral controversy, Engelhardt surveys appeals to intuition, casuistry, consequentialism, hypothetical-choice theory, Rawlsian social good, Habermas's rational discourse, game theory, natural law, and mid-level principlism (e.g., Beauchamp and Childress, *Principles of Biomedical Ethics*).[16] Regardless of the criteria from rival sociological communities or from other communities of empirical scientific investigators—working in the *modus operandi* of appealing to "facts" as decisive regardless of the evaluation of "facts"—the attempts to justify a content-full secular ethics all fall short, and we are left at the precipice of chaos.[17]

13. Engelhardt, *Foundations of Bioethics*, 9.
14. Engelhardt, *Foundations of Bioethics*, 9.
15. Engelhardt, *Foundations of Bioethics*, 33.
16. Engelhardt, *Foundations of Bioethics*, 40–64.
17. Engelhardt, *Foundations of Bioethics*, 42.

> One must appreciate the enormity of the failure of the Enlightenment project of discovering a canonical content-full morality. This failure represents the collapse of the Western philosophical hope to ground the objectivity of morality. This failure bears against theories of justice and accounts of morality generally. It brings all secular bioethics into question. If one cannot justify a particular morality, then one cannot justify claims of immorality. All appears to become a matter of taste. Indeed, if one cannot disclose some lines of conduct as canonically immoral, then the health care provided by Albert Schweitzer and the Nazi death camps will be equally defensible and indefensible. If one cannot discover an objective method to decide when the morally deviant are also the morally wrong, then the action of the morally heinous and the saint will be equally justifiable or lacking in justification, at least in secular terms. One stands on the brink of nihilism.[18]

Engelhardt's exploration of moral ambiguity in secular bioethics leads to a strong conclusion: any attempt to make bioethics more than an "exegesis" of particular moral traditions almost inevitably leads to struggles amidst the will to power. The presumptive strategies listed above for solving moral controversy lack authenticity, for "each presupposes exactly what it seeks to justify: a particular moral content."[19] To this, we might align Wittgenstein's comment: "If someone were to advance theses in philosophy, it would never be possible to debate them, because everyone would agree to them" (*PI*, 128). Within the secular schema that lacks a valuation standard, any imposition of a moral perspective creates "consensus by coercion." Here in the face of this threat, Engelhardt surveys possible ways out of this seemingly inevitable nihilism; moral legitimacy from the secular may exist on the basis of (1) force, (2) conversion of one party to the other's viewpoint, (3) sound rational argument, and (4) agreement.[20] To be clear, appeals to and applications of force—authorized or brute—do not provide *aspirational* answers to moral dilemmas as much as resolutions of one sort or another. Conversion to one's way of thinking or to membership within a moral community (or way of life) is a peaceable goal,

18. Engelhardt, *Foundations of Bioethics*, 65–66.

19. Engelhardt, *Foundations of Bioethics*, 42.

20. Engelhardt, *Foundations of Bioethics*, 67. It is worth remembering that Engelhardt's primary purpose is to describe the actual (and not ideal) state of affairs. In stating force is a "moral legitimacy" within secular bioethics, Engelhardt acknowledges force is not an answer to moral questions, but merely a resolution; hence, "a goal of ethics is to determine when force can be justified" (see also, *Foundations of Bioethics*, 67–74).

and yet Engelhardt is not naïve that such appeals accord naturally within today's secular society.[21] Lastly, as discussed above, sound rational argument offers little hope as a moral perspective—the thing in question—is assumed in drawing moral conclusions, and this leads Engelhardt to the fourth possibility of "agreement." This is an arrangement functioning as "a minimal condition relying on what it is to resolve issues among moral strangers with moral authority: consent. It establishes a secularly acknowledgeable authority for its conclusions: agreement."[22]

Key to Engelhardt's claim that agreement is a secular means is the assumption that permission has no *value*, per se; it is rather an authoritative commitment amongst strangers who share no common values. In place of morals that stem from some shared ontological grounding, this limited curb at the brink of nihilism seeks a sort of administrative parlay without clear moral values. The meeting of two sides in such a scenario can be pictured as two hands shaking in formal agreement on whatever ethical matter at stake. This secularly acknowledgeable authority is a means for peaceable negotiation, and hence Engelhardt sees this ironically as a sort of immanentized, transcendental grounding.[23] It is "a game, a grammatical possibility that cannot be avoided and which comes with inescapable rules but no content."[24] The arguments of *The Foundations*, hence can be recognized as "a disclosure, to borrow a Kantian notion, of a transcendental condition, a necessary condition for the possibility of a general domain of human life and of the life of persons generally . . . this account can be regarded as a transcendental argument to justify a principle of freedom as a side constraint, as a source of authority."[25]

Here, at the edge of modern, moral nihilism, Engelhardt traces in a Kantian and Hegelian spirit[26] a procedural ethic through agreement. How-

21. Engelhardt, *Foundations of Bioethics*, 68.

22. Engelhardt, *Foundations of Bioethics*, 68–69.

23. Appealing to Klaus Hartmann ("On Taking the Transcendental Turn"), Engelhardt explicitly extends Kant's transcendental claims in the area of theoretical knowledge into claims about morality (Engelhardt, *Foundations of Bioethics*, 94–95n82).

24. Engelhardt, *Foundations of Bioethics*, 69.

25. Engelhardt, *Foundations of Bioethics*, 70.

26. Engelhardt defines "transcendental" as "an argument that lays out conditions for the possibility of a major domain of human experience or action. As defining conditions, transcendental arguments are a priori. Here I borrow from Kant, who underscores morality's presupposition of freedom" (Engelhardt, *Foundations of Bioethics*, 94n82). On drawing from Kant and Hegel: "Though my account of the categorical conditions for knowledge and morality has a highly Kantian accent, it can in principle be recast with benefit in more Hegelian terms. A Hegelian categorical account places the argument within the terms of reason and avoids the Kantian difficulty of mediating between the sheer givenness of the object and the predicament of the infinite knower (Engelhardt, *Foundations of Bioethics*, 95n83).

ever thin and threadbare, this procedural ethic, the "principle of permission," functions as a secular moral fabric woven to bring together moral strangers as peaceably as possible. Engelhardt maintains that as with the workings of any scientific reasoning, certain working conditions are assumed as "major, unavoidable elements of the lives of persons." Hence permission and consent do not equate to a grand theory or proposal for content-full virtue but rather have the "character of unpacking a tautology." At the base of this arrangement, the possibility of permission is an "intellectual insight" on what may be disclosed and uncovered as secular morality outside of a "particular proper concrete view of the good life and of content-full moral obligations." [27] In a real sense then, at the heart of Engelhardt's principles of bioethics we can discern a negative structure that follows along the lines not of claim rights but of forbearance.[28] This focus on rights of forbearance further explains Engelhardt's primary principles of permission and beneficence. Permission at its simplest stands as a nonviolent source of authority for secular morality that will not be forced upon someone outside of consent, and beneficence serves as a highly qualified guide for directing the agreement toward securing the best interests of persons. Together, these principles function not as content-rich ideals but as thin procedural steps—or, "chapter headings," "clusters," and "*principia*"—that "indicate the source, beginnings, commencements, or origins of particular areas of the moral life. They are principles in the sense of indicating two different roots for the justification of moral concerns in health care."[29] The principle of permission may be organized under the carefully worded maxim: "do not do unto others that which they would not have done unto them, and do for them that which one has contracted to do." Permission is balanced by the similar maxim for beneficence: "do to others their good."[30]

27. Engelhardt, *Foundations of Bioethics*, 70–71.

28. "The morality that can bind moral strangers hinges on the authority that individuals convey through permission. This morality has a negative structure. It discloses rights and duties of forbearance. The requirement to use individuals only with their consent sets limits. On the one hand, this morality justifies morally content-full joint endeavors through agreements to collaborate. The morality of strangers is the focus of [*The Foundations of Bioethics*]" (Engelhardt, *Foundations of Bioethics*, 102).

29. Engelhardt, *Foundations of Bioethics*, 103.

30. Engelhardt, *Foundations of Bioethics*, 123–24.

Modus Vivendi & the Therapeutic Aims in Wittgenstein and Engelhardt's Critical Mapping

> *Now what makes it difficult for us to take this line of investigation is our craving for generality. This craving for generality is the resultant of a number of tendencies connected with particular philosophical confusions.* (*BB*, 17)

More could be said of Engelhardt's secular, procedural morality, but at this juncture it seems helpful to recall the broader purpose of Engelhardt's methodology. Without such a context, it is likely that the above analysis and framework for secular ethics will appeal only to a few, limited audiences. One such audience could be Kantians and Hegelians, who would appreciate the direction of the argument whereby a certain presumed transcendental reality discloses a condition of life for securing a peaceful existence. Those who reject any application of the dualistic Kantian horizon, however, and seek other strategies or means beyond transcendentals to navigate the immanent and transcendent "spheres" will not find Engelhardt's minimalist paradigm persuasive. Likewise, those who do not agree with a libertarian focus on forbearance will be troubled by permission as the cluster of negative rights that allows persons to coexist in peace. Perhaps most critically, others may question if Engelhardt's presentation of rationality itself is developed enough to function well within his critique of discursive rationality.[31]

Those who favor or oppose these strategies and assertions in Engelhardt will respond accordingly, but for the purposes of this argument it is important to note the *exploratory* aim of *The Foundations of Bioethics*. Not unlike Wittgenstein's explorations with the phenomena of language, Engelhardt explores "the possibility for morally authorized collaboration with moral strangers," which by Engelhardt's own admission may more likely function as a "confession" and "lament" more than a "libertarian manifesto celebrating the value of freedom."[32] Hence these explorations are indeed on one (minimalist) level pursued for what is possible within secular ethics, but the perpetual reminders from Engelhardt on the limitations of these explorations surely direct the focus to something greater. As Wittgenstein noted of solving problems of logic in the *Tractatus*—"how little has been done" (*T*, 28)—so Engelhardt seems to delineate the possibilities within the secular not for the sake of a secular construction but for what the lamentable thinness of the secular indicates. With this pivotal aim in mind, Engelhardt's secular paradigm can be valued for providing a moral *lingua franca* in the midst of

31. See Hauerwas, "Not All Peace Is Peace."

32. Engelhardt, *Foundations of Bioethics*, viii–xii.

diversity—an acknowledgment of "common ground as battle ground"[33]—and, if this reading is correct, it is not too much to say that the strategic maneuverings within the internal logic of the secular are contingent upon—and on a deeper level, inseparable from—the content-full moral meaning that Engelhardt confesses from within traditional Christianity. On this deeper level, the question concerns how far away one may step from one's own ontological commitments, or, how exactly thick or thin is the "sliver of Enlightenment hope" that can be salvaged in Engelhardt's minimalist principles of permission and beneficence? This is a puzzle, for while Engelhardt decries and laments the thin possibilities of the secular, he sees in this "sliver" a possible "limit to the nihilism that challenges us: general canons of secular morality can indeed be articulated. A limit can be given to what appeared to be an unrestricted relativism: the canons of secular morality, although they possess no content, allow for morally authoritative collaboration."[34] These are strong claims, but can such a secular moral collaboration actually exist in the secular apart from Engelhardt's robust ontological commitments? Can the secular essays into "what is possible" function independently of Engelhardt's thick, theological particularity?

In effect, the above question centers on the foundations of the *Foundations,* for it seems clear that the "game" or the "grammatical possibility" in his principle of permission is hardly motivated primarily by a conviction in the certainty of transcendental claims.[35] Rather, Engelhardt navigates amidst the framework of a priori transcendental conditions but with a number of strong qualifications, and it is clear that Engelhardt draws on the "possibility of intersubjective moral coherence and collaboration without presupposing the objectivity of secular morality in the sense of its corresponding to particular content-full moral truths."[36] Here it seems key to note that it is from the *assumption* of seeking moral authority within secular ethics that Engelhardt reasons to content-less moral agreement apart from empirical facts or

33. Engelhardt, "Common Ground as Battleground."

34. Engelhardt, *Foundations of Bioethics,* 83.

35. Hauerwas expresses a similar point: "I admire what I can only describe as the monkish intellectual austerity that governs Engelhardt's development of the contours of his peaceable society. I have my doubts whether he can in fact show that the principle of permission is the 'core' of the morality of mutual respect (Engelhardt, *Foundations of Bioethics,* 117) because such an account seems too close to Kant for someone who has disavowed the Kantian deduction" (Hauerwas, "Not All Peace Is Peace," 39).

36. Engelhardt, *Foundations of Bioethics,* 69, 67–72. See also 94–97, fns 82, 83, and 87) for an example of how Engelhardt qualifies his Kantian position. In particular: "Whatever strengths or weaknesses transcendental arguments may possess in other areas, at least here they offer a means for understanding how the conditions of moral reality are grounded in our character as persons" (97n87).

metaphysical claims.[37] Moreover, within this reading, we may further see Engelhardt operating within a Wittgensteinian mode—where philosophy is employed demonstrably within a greater context, or where it functions in a contingent or therapeutic service.[38] Wittgenstein and Engelhardt both employ a method of philosophizing that is conditional and not aspirational. Both pursue philosophy as a therapeutic cure (*PI*, 133)—or where philosophy addresses questions "like an illness" (*PI*, 255)—aimed at disarming scientific theorizing, and what is embedded within such a philosophical method is the assumption of a content-less philosophy. Such a content-less philosophy, moreover, is an inside-out maneuver that reverses secular logic against itself, rendering its own claims for objective theories or norms vacuous. Though Engelhardt does intend to offer some tools of consensus and a limited moral *lingua franca*, his framework demonstrates on a deeper level the staunch limitations of secular ethics and bioethics. Within secular theoretical thinking, the best one can hope for is some extension of content-less, transcendental conditions whereby moral strangers can agree to form an intersubjectival ethic of non (or "lesser") violence. To examine further this common spirit of philosophy, it is helpful now to examine two primary parallels between Wittgenstein and Engelhardt: (1) their use of philosophical investigations as a heuristic or as a pedagogical exercise that enables us to better grasp (2) the parallel between Engelhardt's *modus vivendi* and Wittgenstein's forms of life (*Lebensformen*).

37. Engelhardt, *Foundations of Bioethics*, 95n82.

38. Like Engelhardt, Wittgenstein challenged the dominant theoretical paradigm of his time, and as a result he, too, was frequently misunderstood. He once spoke of this disconnect, commenting that one of his works, *Philosophical Remarks*, "is written for such men as are in sympathy with its spirit. This spirit is different from the one which informs the vast stream of European and American civilization in which all of us stand" (*PR*, 7). There are layers and conditions involved in reading Engelhardt and Wittgenstein, and both must be approached in a certain *spirit*. Edwards speaks of this spirit in Wittgenstein as "prophetical and evangelical," both terms that may be ascribed as well to Engelhardt. Moreover, to grasp Wittgenstein and Engelhardt's therapeutic similarities, we should recall the need for a metareading of them both. As has been argued above, to read the early apart from the later Wittgenstein is to misunderstand enormously important developments in his methodology and thinking. At the same time, to ignore the bridge and commonality of probing the mystical between the two Wittgensteins can lead to polarized notions that the later Wittgenstein explicitly contradicts the earlier. A more nuanced metareading is required. Similarly, those who highlight only Engelhardt's libertarian aspects will come away thinking that his principles of permission and beneficence are the main points or purpose of his work. This is a mistake on both accounts when we look to the way that both Wittgenstein and Engelhardt explore essential conditions of immanent reality as something akin to pedagogical exercises for the sake of encountering a greater transcendent reality.

Regarding the first parallel, it is helpful to note further the therapeutic aim for Wittgenstein's grammatical investigations and Engelhardt's analysis of bioethics. As intimated above, Engelhardt's minimalist and lamentable procedural ethic stands *to do something* in an age of demoralized and deflated morality, and what it *does* has a two-sided ambition both to produce some procedural possibility for peace and to direct one to imagine greater transcendent possibilities beyond the sphere of the immanent and finite. Within the secular investigation of permission (*The Foundations of Bioethics* 2nd ed., 1996) and the cultural narrative explaining our postmodern moral predicament (*After God*, 2017), Engelhardt follows a heuristic technique of examining the possibility for a consistent bioethical methodology before turning to broader philosophical and theological matters. Conclusions on bioethical issues certainly matter, but a deep, pedagogical theme within Engelhardt's work is to investigate even more the possible groundwork for morality within the assumptions of postmodernity.

Here the family resemblance with Wittgenstein is strong, for he, too, seeks through grammatical investigation first to uncover the possibilities—within our theorizing and scientific age—for knowing as a heuristic exercise that directs one toward "what is higher" (*das Mystiche*). Wittgenstein also cares deeply for final conclusions even while he insists first on examining possible methodologies—through logic and language—for knowing. These two poles are key for understanding Wittgenstein, and as Edwards has noted, the *Tractatus* is a book of "radically divided sensibility" between Wittgenstein's technical explorations with logic and his appeals to step beyond that pedagogical ladder toward "what is most important" and "what is higher" in *das Mystiche*.[39] Certainly, we must note the clear difference between Wittgenstein's aphoristic gestures toward *das Mystiche* and Engelhardt's explicit confession in Jesus Christ. Wittgenstein takes on the exercises of "rationality as representation" in the *Tractatus* and is reduced to a nameless mysticism given the sharp lines he has drawn between "rational" thought and "mystical" will. In contrast, Engelhardt issues the radically Christian call for moral knowing through the ascetic life of putting on Christ (Rom 13:14), and this is a disciplined, ecclesial, prayerful, liturgical, and iconic way of life. Much more could be said of these critical differences, and yet the point for now is to identify the common spirit of philosophy in Wittgenstein and Engelhardt.

Moreover, in addressing this relationship between philosophical thinking and "what is higher," it is important to note that both philosophers place an emphasis on philosophical investigations that are *suspended* from mystical conclusions, and this *suspension* is a mark against those wishing

39. Edwards, *Ethics without Philosophy*, 7.

to place (finally) Wittgenstein and Engelhardt's philosophical exercises on a divided plane or Kantian horizon. Rather, the relationship between their philosophical exercises and "what is higher" is hierarchical and more dynamic than some hard divide to be spanned via Kantian transcendentals. Engelhardt and Wittgenstein rather examine bioethics and logic/language respectively as pedagogical tools that should direct one above toward *transcendence* (or, for Engelhardt, to know the transcendent God in Three Persons), but neither seeks a higher plane of *transcendentals*. In this way, Engelhardt's permission follows a Kantian (and Hegelian) spirit, but not for the sake of furthering permission as a transcendental or as a dialectical rationality. Similarly, as argued above, while Wittgenstein's thinking matured notably over the course of his life, there is a unity between the "early" and "later" Wittgenstein(s) in that all his writings, including the *Tractatus*, were concerned with philosophy as a formative tool that clarifies and humbles human attempts at discursive rationality to extend beyond what it can do. Speaking of Wittgenstein's "Lecture on Ethics," Edwards describes an apophatic form within Wittgenstein's ethical thinking:

> The task of ethics is the restoration of meaning to my problematic world, a meaning that lies outside the world, in the philosophical self which is my world's limit . . . [this ethical] sense of life cannot be talked about, nor can it be said what exactly is the nature of good willing that makes it clear. These things belong to the realm of *das Mystiche*: they *show* themselves. The *Tractatus*, by demonstrating clearly what can and cannot be said, makes it clear how impotent thought (language) is when it comes to the really important things in life. Silence, the silence in which *das Mystiche* shows itself is the paradigmatic response of the enlightened philosopher who has kicked away the ladder of philosophy after he has climbed up it.[40]

Edwards proceeds to highlight the importance of Wittgenstein's "doctrine of showing" to reinforce that the "unsayable" is communicated not through an "intellectualist conception of rationality" but through other modes such as the willing self, again reminiscent of Kierkegaard. In short, what emerges from both is a therapeutic distinction between immanent philosophical investigations (or exercises) and other modes of knowing that embrace transcendence, and this in turn opens the door to mystical methods of knowing beyond discursive rationality or theorizing. Neither Wittgenstein nor Engelhardt employs philosophy as a constructive, systematic way for authentic knowing but rather as a critical exercise in mapping

40. Edwards, *Ethics without Philosophy*, 49 (emphasis original).

the thin possibilities that can be discerned within the immanent systems of logic/language and bioethics.

With this heuristic and therapeutic methodology in mind, we are better able to understand Engelhardt's frequent appeals to *modus vivendi*, particularly in *After God.* First and foremost, we should note how *modus vivendi* accords within Engelhardt's critique of immanentized bioethics. *After God* traces a narrative of secular morality as a deflated endeavor grounded solely within the rights and wishes of individual sovereigns, and hence morality departs from any substantive vision of harmonious communal life or bodily corporate unity and exists solely within an individual's lifestyle choice. Political structures framing the morality of the *polis* become mere arrangements, a *modus vivendi* within which one is "willing to live for at least the time being."[41] This *modus vivendi* possesses a law and public policy that reflect "one among a plurality of freestanding accounts" of morality.[42] In short, within the strict limits of secular morality after God, *modus vivendi* emerges as a mere "political life-style choice," and the "authority" of the political state exists within the machinations of the market. Following Buchanan and Tullock (*The Calculus of Consent*, 1965), Engelhardt points out that when "enough" of a consensus surfaces, a particular political order arises not as a contract or other reasonably clear agreement but as a "temporary resting point accepted by enough persons so as to be stable." Hence a population (as "we") agrees "as long as there are enough to impose an 'agreement,' a *modus vivendi.*"[43]

Engelhardt also points to George Tsai's definition of a *modus vivendi* as:

> . . . a strategic compromise among contending groups in a society none of whom is in a position to impose its preferred way of life on the others without unacceptable costs and each of whom as a result adopts a policy of mutual accommodation as the best that it can hope to achieve under its circumstance.[44]

Even beyond Tsai, however, Engelhardt sees the possibility of a *modus vivendi* where subjects of a state do not have genuine interests in the state's continued functioning. The motivations and realities facing citizens may be complicated to a point where they await the opportunity for a regime change, even a civil war. Clearly, this understanding of *modus vivendi* underscores

41. Engelhardt, *After God*, 18.
42. Engelhardt, *After God*, 20.
43. Engelhardt, *After God*, 192.
44. Tsai, "Lamentable Necessities," 794. See also Engelhardt, *After God*, 193.

the minimalist nature of such a politic, and yet this is the reality that Engelhardt admits in an age where moral differences are as stark as they are. The implication of Engelhardt's *modus vivendi* is that subjects who recognize the true nature of postmodern moral diversity may often face a complicated calculus of citizenship where "one can prudently support constitutions that can constrain the tyranny of the majority, not merely the tyranny of the few." One can easily envision scenarios where one must distinguish not between ideal options but "among more or less onerous more-than-minimal states." In short, *modus vivendi* is a wake-up call for "the proponents of liberal agendas of equality, human rights, and human dignity that often resolutely hold 'these truths' to be either self-evident or rationally established." In fact, such liberal optimism falls radically short.

> They fail to recognize the socio-historically conditioned character of their claims. Bioethicists act as if they could establish moral and bioethical truth and as if they were experts regarding that truth. The view that bioethics is grounded in a common human morality is core to Beauchamp and Childress's highly influential textbook (*Principles of Bioethics*, 1979). This textbook, throughout its many editions, makes such claims. . .[45]

Far from any aspirational political structure, the Western democracies typically appealed to in bioethics do not operate upon any agreed-upon, canonical rationality. That Enlightenment dream has passed. Rather, implicitly or explicitly, bioethics exists within an inevitably political space that is all too often motivated and "encouraged" through the "seductive rhetoric of mass media democracies." What remains is a sociopolitical struggle not over ideals that are first-order theories; the struggle is much less a content-full engagement of moral visions but rather of sophistry and politics. *Modus vivendi* in this sense reveals the deep veins of biopolitics at work within bioethics:

> At the most, there can be empirical judgments, educated guesses, regarding the likely political stability of some versus other forms of *modus vivendi*. The culture wars now also reflect disputes regarding the acceptability of a particular form of governance as a *modus vivendi*.[46]

Rather than search for some theoretical fault or some explanation of where things went wrong through a logical or rational misstep, Engelhardt is honest in identifying the tangible agents of change employed in concrete,

45. Engelhardt, *After God*, 195.

46. Engelhardt, *After God*, 195.

everyday life. Engelhardt's *modus vivendi* hence functions as a key tool of analyzing the way things actually are in bioethics. As with Wittgenstein's later focus particularly on language as it is oriented within concrete, everyday life rather than stemming from "philosophical thought" (*PI*, 298), so Engelhardt looks to the actual conditions that affect bioethics more than discursive reason. Such a posture, moreover, agrees with Koch's critique of cognitivist and principlist approaches that seek to address bedside dilemmas through philosophical thinking. No one, Koch argues, actually brings Kant's *Critique of Pure Reason* to a clinical consultation.[47]

As another way to understand the trend in bioethics to look "internally" or within "principles" for the way forward—and hence the challenge and relevance of Engelhardt's *modus vivendi* and Wittgenstein's *Lebensformen*—we should remember the longstanding tendencies in traditional Western philosophy to exist with the metaphysical "regime of thinking." Wittgenstein's later work specifically challenges this metaphysical tendency that Kerr aligns with notions of gnostic dualism and Cartesianism, where the human ideal arises above the body, presenting a mentalism that Wittgenstein ardently denied. Kerr remarks that Wittgenstein's therapy is essential to helping free us from tendencies to look "within" and rather to note the historical and social conditions that reacquaint us with embodiment. Hence "Wittgenstein's later writings are key texts in subverting the entire metaphysical tradition which is constituted by rancor against the physical and historical conditions of human life."[48]

Wittgenstein and Engelhardt both should be viewed as contesting exactly this metaphysical tradition, and both pursue their respective analysis of the roots of language within *Lebensformen* and of politicocultural *modus vivendi* for the sake of uncovering what is genuinely at work beyond the troubled language of principlism and theoretical philosophy. The point is not to discount entirely principles (for Engelhardt) or language (for Wittgenstein) as much as the goal is to recount the inevitable connections between these discursive activities and the facts of living that are commonly obscured within the logic of the secular. Hence Savicky notes that Wittgenstein's grammatical investigations bring to attention "the fact that the speaking of language is a part of a form of life, and imagining a language means imagining a form of life (PI, 23, 19)." In other words, for Wittgenstein "forms of life (or facts of living) are what have to be accepted. They are the given."[49] Wittgenstein furthers this point by stating:

47. Koch, *Thieves of Virtue.*

48. Kerr, *Theology after Wittgenstein*, 188.

49. Savickey, *Wittgenstein's Art of Investigation*, 112.

> "So what you are saying is that human agreement decides what is true and false?"—What is true or false is what humans say; and it is in their language that human beings agree. This is agreement not in opinions, but rather in form of life. (PI, 241)

To this we can align Engelhardt's principles of permission and beneficence that function as *principia*. These *principia* do not uncover some pure, secular, crystalline position within bioethics derived through "rational argument or common belief," but rather *principia* function as "the minimum grammar for secular moral discourse."[50]

So where does this leave us? Admittedly, some may object to the possibility of Engelhardt salvaging anything from the project of secular morality, for such a move seems to grant some justification or rationality for (some) ethics from within the secular. If the ontology of the secular, after all, leads only to a politics of power and violence, how can we expect any desirable bioethics to emerge from the ruins? A similarity here may be drawn with Wittgenstein's critique of language that maintains a form of consensus through communal language games. In a theological way, Wittgenstein seems to bracket some transcendent meaning—even the existence of God—as a way of justifying communal language, and one could proceed to think even that Wittgenstein falls prey to his own charge of metaphysics. Such a concern for Wittgenstein would indeed be incisive for plumbing the meaning and reality below the surface methodology Wittgenstein employs. Yet it is the nature of *methodology* in particular that seems to answer such a concern for Wittgenstein bracketing God or some other transcendental meaning. I have argued earlier that Wittgenstein notably emphasizes methodology, and this is a focus and task that inherently limits the scope of his overall work. Simply put, Wittgenstein will not be satisfying for those seeking to go beyond methodological clarity and into philosophical speculation. Hence in a sense, it is right to think Wittgenstein brackets God in his philosophical method without ever really unpacking that grounding assumption. Equally, however, we should note that Wittgenstein does *address* this bracketed grounding for his methodology even if in a mysterious way evidenced in the concluding *aporias* of the *Tractatus* and throughout the *Investigations*. Wittgenstein declines to speak philosophically or theoretically of that which surpasses speech, and his focus with philosophy hence pursues the practice of clarity. For Wittgenstein, "the work of the philosopher consists in assembling reminders for a particular purpose" (*PI*, 127).

Here Engelhardt follows a similar path. As Wittgenstein traced the phenomena of language for a particular purpose within a conditional

50. Engelhardt, *Foundations of Bioethics*, 123.

landscape, Engelhardt engages the possibilities for bioethics from within a landscape of postmodern, pluralistic skepticism that characterizes contemporary Western societies. Unlike Wittgenstein, however, Engelhardt is far more explicit in identifying the limits of such a purpose; the current task of ethics and bioethics within the contemporary public square of Western society is often far from conversations about the good or even conversations amongst authentic moral communities—be they traditional Confucian, Jewish, Muslim, or Christian communities—about what is the good life. Without God, there is no ultimate good to be found anywhere, and hence Engelhardt does not celebrate the limited consensus of permission amidst "the chaos, or even much of the diversity, and surely not the moral perversity and vacuity of this landscape." Instead, Engelhardt seeks a "secular means for coming to terms with the chaos and diversity of postmodernity. The means are meager and offer no transcendent fulfillment. But they are all that is available in general secular terms."[51]

In short, Wittgenstein and Engelhardt both present an *aporia* of simultaneously employing a critique of secularism while nevertheless engaging in the available possibilities of the secular. Doubtless this is a puzzle, and for some the puzzle seems something of an inevitability for life in the wake of Enlightenment modernity.[52] Contrary to those who may appeal to a utilitarian argument or some version of substantive liberalism by tracing the methodology of language and the principle of permission in a secular society, Wittgenstein and Engelhardt seek a particular purpose in the engagement of immanent language and permissive means. Considering the phenomenological side of Wittgenstein, language stands to reveal something of the Ethical (*LE*, 43–44) and to serve as an ascetic practice that forms the person to see the world rightly, looking through the deflation of theoretical metaphysics toward *das Mystische*. Considering the pragmatic side of Engelhardt, admitting difference in bioethics amidst postmodernity stands to expose the limits of any secular good for all its vacuity and dissatisfying immanence, hopefully leading one to long for something with more transcendent meaning.

51. Engelhardt, *Foundations of Bioethics*, 10.

52. Ruping Fan argues that the complete lack of "substantive conclusions" in any version of "substantive liberalism" renders "an Engelhardtian non-substantive (or procedural) liberalism . . . morally unavoidable" (Fan, "Unjustifiability of Substantive Liberalisms," 233).

Seeking Conversion: Discourse Amidst the Secular and Christian

> *If one wants more than secular reason can disclose—and one should want more—then one should join a religion and be careful to choose the right one.* —Engelhardt[53]

Beyond this staunch critique of secular bioethics, however, some have questioned what hopes remain for any non-Christian discourse within the post-Christian public square. Has Engelhardt's Kantian-Hegelian critique left us with moral silos that cannot speak with one another amidst any shared discursive grounding? We turn now to this question of reasoning—in both a Christian and a non-Christian idiom—in the following with a twofold goal: first, to identify a few particular critiques from those who think Engelhardt has overly limited the possibilities for reasonable discourse amidst differing moral communities. The second goal takes us beyond Engelhardt's critical analysis of secular bioethics to understand the possibilities for discourse in the public square from Engelhardt's perspective as a traditional Christian. This second goal brings us to Engelhardt's nuanced consideration of natural law and the possibilities for shared language and idioms amidst staunch moral strangers.

Turning to those who challenge Engelhardt's strict limits on discourse amidst differing communities, we may consider Lucie-Smith's narrative analysis that he proposes for reading Engelhardt. As Lucie-Smith details, narrative is bound together with a choice to start not with abstract, deductive principles but from within a particular story that is more inductive in nature—inductive in that one can trace the meaningful account of reality through and in consideration of a community's traditions, mythology, and rituals. In contrast, starting with abstract first principles assumed to be universal leads to a more scientific, theoretical model, and this is the Enlightenment approach for justifying morality that Engelhardt argues is fundamentally flawed.[54] Lucie-Smith summarizes Engelhardt's narrative methodology:

> Thus what often presents itself as a purely philosophical view (one seemingly backed by abstract argument) is nothing of the sort, but rather the reflection of the person who holds his position as a member of this or that community, this or that tradition, and a product of their particular lived history. In other words, belief is something that comes to us as narrative, not as

53. Engelhardt, *Foundations of Bioethics*, xi.

54. Lucie-Smith, *Narrative Theology, Moral Theology*, 1, 47–72.

> something that can be sustained through seemingly universal rational categories . . . [hence every moral statement] is not ever merely a statement: it carries with it a commitment to a certain way of life . . . a vision about the whole of life. The moral statement is the tip of the iceberg, and what lies beneath the surface is the whole story, perhaps a whole raft of views on family life, God, the role of women and Church discipline and authority, in other words, a narrative.[55]

Narrative for Engelhardt plays out in at least two ways: first, Engelhardt denies any such thing as a freestanding statement or claim to impartial and abstract reasoning. Narrative contexts are undeniable, and hence any philosophical maneuvers seeking autonomous universals and rationally discoverable content-full morality are fundamentally flawed. Secondly, Engelhardt grasps the reality of his own narrative.[56] Engelhardt explicates his distinctively religious "view from somewhere"—namely, as a "Texan Orthodox Catholic."[57] The "Texan" qualifier may seem remote from Engelhardt's claims on the necessity of a particular moral narrative, and yet his dry-humored Texan roots seem to function as a heuristic more than an eccentric or rhetorical style. One's place, era, and heritage are deeply formative and matter more than many academic philosophers may be willing to admit, and in this spirit Engelhardt's persistent reminders of "being Texan" seem to function intentionally as a thorny prod in the side of liberal "cosmopolitan ecumenists."[58]

55. Lucie-Smith, *Narrative Theology, Moral Theology*, 49. See also Engelhardt, *Foundations of Bioethics*, 14.

56. Engelhardt interweaves his personal narrative with his philosophical and theological arguments: "If one wants more than secular reason can disclose—and one should want more—then one should join a religion and be careful to choose the right one. . . . I indeed affirm the canonical, concrete moral narrative, but realize it cannot be given by reason, only by grace. I am, after all, a born-again Texan Orthodox Christian, a convert by choice and conviction, through grace and in repentance for sins innumerable. . . . My moral perspective does not lack content. I am of the firm conviction that, save for God's mercy, those who willfully engage in much that a peaceable, fully secular state will permit (e.g., euthanasia and direct abortion on demand) stand in danger of hell's eternal fires" (Engelhardt, *Foundations of Christian Bioethics*, xi).

57. Engelhardt, *Foundations of Bioethics*, xi.

58. Engelhardt largely follows MacIntyre's use of the term cosmopolitan in *After Virtue*. Cosmopolitan ecumenists are "rootless and eclectic individuals" (Wildes, "Engelhardt's Communitarian Ethics," 89) who believe strongly in the possibility of secular reasoning while denying the real diversity among moral perspectives. "Such cosmopolitan ecumenists hold that men and women share enough in common so that a concrete and authoritative moral consensus can be discovered in societal undertakings that will allow them to justify a particular bioethics and to direct health care polity with moral authority. . . in addition, cosmopolitans often live their lives far from the substantive moral convictions that guide committed Orthodox Jews, Orthodox Catholics,

Personalities, character, and the virtue of persons matter. Picking up on this and noting Wittgensteinian overtones, Nash comments:

> We humans do not share a common rational human ethic, but instead we behave and have proclivities due to a variety of factors. Engelhardt's exhibition of *Lebensform* also provides a corrective for overly progressive Christians who try to assume that there is a standpoint from nowhere. This cannot be conceded. After all, Christ was born in Palestine, and not in Palestine, Texas.[59]

Through reminders of Engelhardt's Texan roots or through his narratival critique of crystalline Enlightenment rationality, Engelhardt does not veil the tradition and narrative that underscores his moral vision. To clarify, this is not to say that Engelhardt presumptively champions his narrative as though others simply need to acknowledge its superiority over other narratives on an equal playing field. The narrative approach Engelhardt holds to is not a reissuing of metacriteria to be evaluated by autonomous rationalities as though one were seeking to redo the Enlightenment process, seeking *the* clear rational choice. Advancing narrative in such a naïve way misses the corrective and clarifying intent to caution against presumptive theorizing, seeking a view from nowhere. However, moving from this qualification, Lucie-Smith is concerned that Engelhardt overemphasizes the gulf between the narrative of a content-full moral community—which Engelhardt labels a "moral enclave"[60]—and the secular sphere. Lucie-Smith points to MacIntyre and Hauerwas as two others who emphasize the importance of narrative, community, and tradition, and yet, for Lucie-Smith, Engelhardt mistakenly turns to a mystical, liturgical interaction between the two spheres rather than following MacIntyre's call for "necessary dialogue" in their midst.[61] This critique stems from Lucie-Smith's understanding of Engelhardt as denying "intrinsic rationality or irrationality" in some acts. If there is no genuine bridge between moral enclaves, if each is a "closed fortress," we are left to MacIntyre's blunt choice between Aristotle's virtue and Nietzsche's nihilism, with Engelhardt falling

Roman Catholics, Protestants, Moslems, and others who understand themselves within enduring and concrete moral traditions. It is difficult for cosmopolitan ecumenists to understand life within the embrace of traditional communities framed by transcendent commitments or to fathom the gulfs that separate different communities of the ideologically committed (Engelhardt, *Foundations of Bioethics*, ix).

59. Nash, "H. T. Engelhardt, Jr.," 273–74.

60. Engelhardt, *Foundations of Bioethics*, 82.

61. Lucie-Smith, *Narrative Theology, Moral Theology*, 54.

into the latter.[62] Hence for Lucie-Smith, Engelhardt's narrative falls into a dualistic trap as (1) the failure of the Enlightenment project leaves us with a content-less procedural ethic and (2) content-full ethics exist only within moral enclaves:

> If we are to accept narrative as the fundamental category in ethics and moral theology, and by extension the idea that narrative is only nurtured in specific communities and traditions, then, it seems, we are abandoning any idea of morality as being universally valid. If the Enlightenment Project has failed (and it may well have done so) are we forced to admit the uselessness of discursive rational argument? Engelhardt, perhaps more clearly than MacIntyre and Hauerwas, shows us the consequences of doing so.[63]

What is clear is that Lucie-Smith reads Engelhardt's narrative as being advanced in "contradistinction to reason, not in harmony with it; narrative as a replacement to failed reason, not as a help to a still viable reason."[64] Seeking a peaceable secular grounding of some sort, Engelhardt's principle of permission is chosen through a will to morality not arrived at through any rational process. The "bogey of coercion" is hence replaced by the "specter of self-coercion," and the possibility of moral friendship seems far from reality.[65] Lucie-Smith's assessment of Engelhardt is persuasive on a number of points, and yet what is not included in Lucie-Smith's challenge is a critical distinction between modes of knowledge. The discursive rationality that Engelhardt sees at work within bioethics as an Enlightenment offspring is but one mode of rationality that is engaged in Engelhardt's work. Lucie-Smith is correct to identify the narrative nature of Engelhardt, and yet to critique the limits of discourse in Engelhardt's principle of permission, Lucie-Smith should at least engage Engelhardt's hierarchy of rationality.[66] Noetic rationality in particular remains as a persistent theme throughout Engelhardt's work such that it is not helpful simply to point out that Engelhardt's *modus vivendi* harms discourse, for there are multiple modes or layers of possible discourse assumed in Engelhardt beyond the monolithic discursive rationality of the Enlightenment.

62. Lucie-Smith, *Narrative Theology, Moral Theology*, 56.

63. Lucie-Smith, *Narrative Theology, Moral Theology*, 60.

64. Lucie-Smith, *Narrative Theology, Moral Theology*, 72.

65. Lucie-Smith, *Narrative Theology, Moral Theology*, 69.

66. As discussed below, Goss and Vitz ("Natural Law among Moral Strangers") exemplify such a nuanced critique of Engelhardt that appreciates his distinctions on modes of knowing.

Similarly, Kevin Wildes reads Engelhardt in terms of the relationship between different communities, and yet he challenges the notion that communities should be marked by the differences or space between themselves.[67] Engelhardt is a "communitarian" thinker who navigates the postmodern condition through a political framework centered on the distance between moral friends and moral strangers,[68] and, as intimated above, this is a mapping strategy, an attempt to understand the valueless procedures that are possible between communities that possess different visions of content, authority, and how ideally to relate to other communities. Wildes has no issue accepting Engelhardt's clarifying focus on community narratives, but rather than seeing the space between communities as antagonistic he suggests moral similarities may exist. In other words, rather than classifying the postmodern populace into moral friends and strangers, there are "moral acquaintances" who may work together.[69] Doubtless, Engelhardt would counter that Wildes has largely underestimated the nature of the gap between communities. When moral strangers share moral similarities, as Wildes suggests, the question remains *why* these similarities are shared, and when the answer to that question also acknowledges sincere metaphysical differences, the relationship of moral acquaintances as a qualification for the moral friends-strangers paradigm is challenged. Engelhardt admits the fact of some shared moral visions and common premises as "loyalties" are indeed "divided among different moral communities,"[70] and this explains within his framework the occasions when moral strangers seem closer than their rival foundations may imply. Exceptions granted, Engelhardt's focus on particular, substantive, moral communities within the dominant secular cosmopolitan culture makes sense as he points, for example, to the reality of life within the apostolic Christian community where the "Epistle to Diognetus" records that Christians "dwell in their own fatherlands, but as if sojourners in them; they share all things as citizens and suffer all things as strangers. Every foreign country is their fatherland, and every fatherland is a foreign country."[71] In

67. Wildes, "Engelhardt's Communitarian Thinking," 77–78, 91.

68. Wildes, "Completing the Picture," 100–01. Wildes suggests three models of communitarian thinking: the exclusive community that sees moral knowledge as rooted in the community itself such that "morality is simply part of its way of life"; the inclusive community which aligns with Karl Rahner's "anonymous Christian" where the religious truth of the community is a public truth for all; and the pluralistic community which offers no hierarchy but supposes all communities may equally learn from others. Wildes sees Engelhardt as mapping the landscape of postmodern bioethics all the while having an exclusive community model in mind.

69. Wildes, "Engelhardt's Communitarian Thinking," 90–91.

70. Engelhardt, "Re-Reading *Re-Reading Engelhardt*," 295.

71. "Epistle to Diognetus" as quoted in Engelhardt, "Re-Reading Re-*Reading*

light of such substantive moral difference, Wildes's reading seems to offer more of a subtle nuancing than a critique.

A related and more significant question for Engelhardt's principle of permission, however, is formulated by Griffin Trotter.[72] Trotter has no problem accepting the majority of Engelhardt's assessment regarding the theoretical foundations of secular bioethics, but he argues that the space between moral friends and strangers can be a middle ground for a discursive "intermediate" bioethics. Intermediate bioethics follows the assumption that any attempt to rule out discursive inquiry is simply inexpedient, and that a discursive compromise is possible between religiously thick traditions and communities such as Buddhists and traditional Christians. "Intermediate bioethics is an instrumental ethics that mediates moral controversies between divergent moral communities and aims at peaceful coexistence" primarily by extending Engelhardt's principle of permission from the action of consent into the sphere of agreeing to rules and practices that will govern relations between individuals and communities in a normative mode of generating content.[73] In short, Trotter further specifies Wildes, and both can be seen questioning the nature of reasoning and speculation amidst metaphysically and morally distinct communities. Trotter advances an understanding of reason that seeks to assimilate noetic theology and experience, and it is interesting that he compares hypotheses derived from *noesis* to other hypotheses, including scientific hypotheses, that are "reproducible" and hence observed to follow some type of ordered, natural paths. Further, in addition to discursive reasoning from within the Christian community sharing properties with scientific thinking, Trotter posits that traditional Christian knowledge needs to be further developed into categories such as bioethics. This may seem obvious, and yet Trotter's point is that to move from Christian theology/doctrine to Christian bioethics is to move within a discursive rationality that is not relegated, as Engelhardt maintains, solely to the moral enclave of traditional Christianity.[74]

Here we can summarize a challenge to Engelhardt's position on the possibility of moral discourse amongst moral strangers. Taken together, Lucie-Smith, Wildes, and Trotter all appreciate Engelhardt's critique of the grounding for secular, theoretical bioethics, and yet they maintain that his critique more or less runs aground through a fundamentalist dualism regarding discursive rationality. They share a common critique that

Engelhardt," 301. See also Grant and Graham, *Apostolic Fathers,* 2:359, 2:361.

72. Trotter, "Is Discursive Christian Bioethics an Oxymoron?," 203–27.

73. Trotter, "Is Discursive Christian Bioethics an Oxymoron?," 209, 207–10.

74. Trotter, "Is Discursive Christian Bioethics an Oxymoron?," 210–17.

Engelhardt has taken a helpful insight on the nature of nonsecular morality embedded within communities and narratives too far and mistakenly ruled out hope for any substantive common morality via some degree of shared discursive rationality. Or, at the least, he has stopped short by seeing the principle of permission as a valueless procedure that should extend into a middle ground of intermediate bioethics amidst moral acquaintances. Lucie-Smith, Wildes, and Trotter agree with Engelhardt that the fullness of truth is embodied within the Christian community, and this claim entails an exclusivity that is scandalous (I Cor 1:23).[75] What they contest in Engelhardt, however, is the way that he has overly limited the possibilities for rational discourse outside of the Christian community. Seemingly, Engelhardt has removed any possibility for non-Christian moral idioms in the public square, hence relegating bioethics into a segregated, nondiscursive system of communes that never connect outside of the thin constructs of consent/permission detailed above.

This reading of Engelhardt seems confirmed by Engelhardt's "Recent History of Christian Bioethics" (2014) where he decries particularly the early Christian theologians-turned-bioethicists who sought to function well in "an increasingly secular context" by employing a post-Christian discourse. As noted in chapter one, these theologians sought to present "Christian bioethics within an idiom they hoped would be open to all," hence accepting a morality from the humanities that would substitute for God.[76] Roman Catholic theologians such as McCormick arguably most exemplify this collapse of traditional Christian language by accepting the "conceits of natural law" and aspiring to establish "moral views" through what was deemed to be "purely philosophical arguments."[77] Seemingly, Engelhardt's point is that any common discourse amongst moral strangers lacking a common moral vision is a meaningless

75. In differentiating the ways of securing moral content within traditional Christianity from that of the secular mind, Engelhardt freely admits the scandal of his particularist claims. Christian and secular bioethics vary essentially, and this particularity is "troublesome" for the secular mind. "It has a sectarian character offensive to an ecumenical spirit. The personal character of the Christian narrative, which is attractive to many, is for this reason also disturbing. Traditional Christianity promises to locate all moral concerns, all encounters with moral and natural evil, within a very particular history of creation, the fall, and redemption. Though this account does not avoid any of the particulars of the human drama, it excludes competing non-Christian understandings, which may be similar but not the same. The very thickness of Christianity's moral power is tied to its exclusiveness. . . how can a bioethics be Christian unless it differs at least in some respect from a secular bioethics, unless it is in some sense sectarian?" (Engelhardt, *Foundations of Christian Bioethics*, 1).

76. Engelhardt, "Recent History," 152.

77. Engelhardt, "Recent History," 149.

endeavor. Boaz Goss and Rico Vitz, however, disagree with such a reading of Engelhardt and suggest a more nuanced understanding.

Goss and Vitz point out that Engelhardt himself engages in a broad array of "acceptable kinds of moral discourse," and they point to Engelhardt's endorsement of "therapeutic deceit" as one such example.[78] When a physician purposefully lies to his patient in certain circumstances to achieve the necessary therapeutic treatment, such a "meritorious" or even "obligatory" discourse is certainly flexible and adapted toward seeking a common ground. Similarly, Goss and Vitz point to other instances where Engelhardt appeals to the "concept of a person" as a reasonable grounding for understanding medicine, and they conclude that it is mistaken to see Engelhardt's critique of secular bioethics as a simplistic call to shun ethical dialogue outside a community of moral friends. To clarify Engelhardt's true point, Goss and Vitz discuss the twofold manner in which Engelhardt addresses natural law. On the one hand, Engelhardt is strikingly critical of natural law "as understood by the Stoics or later by the scholasticism of the West: an impersonal legislation, discursively available."[79] Such a natural law is not acknowledged explicitly as a "gift" to the human heart "coeval with creation" but rather as an epistemology grounded in "human inclinations or biological functions." This is a natural law apart from God where medical ethics can be discerned primarily by looking to biology and nature itself,[80] and this natural law can even be viewed as an "independent moral framework," something beyond God that constrains both creation and the Creator.[81] Goss and Vitz, moreover, carefully connect this understanding of natural law to later scholastic philosophers such as Grotius and not to Aquinas, for on Aquinas's account, "natural law is fundamentally grounded in the eternal law," whereas for Grotius and others, the natural law is "fundamentally grounded in human nature."[82] This move to present an account of natural law apart from God is consistent with the New Natural Law Theory popularized by Germain Grisez, John Finnis, and others in the 1960s, coincidentally or not around the same time that bioethics was emerging in academic and social spheres. Goss and Vitz hence draw

78. Goss and Vitz, "Natural Law among Moral Strangers," 284.

79. Engelhardt, *Foundations of Christian Bioethics*, 172.

80. Engelhardt points to McFadden, *Medical Ethics*, as an example of medical ethics derived from "biological or physicalist" guidelines. Similarly, he points to Kant's moral rationality in *Metaphysical Principles of Virtue* that offers detailed moral constraints on "matters such as masturbation, the sale of body parts, and smallpox inoculation" (Engelhardt, *Foundations of Christian Bioethics*, 218–19).

81. Engelhardt, *Foundations of Christian Bioethics*, 173.

82. Goss and Vitz, "Natural Law among Moral Strangers," 286–87.

out that the New Natural Law Theory articulates an account of natural law that brackets religion as a basic human good, but this human good is "attenuated" in that it pursues thin questions of natural theology addressing the order and freedom of humanity within the cosmos. Such a natural law amounts to "a kind of Kantian postulate about transcendent harmony" or perhaps a "Sartean postulate about transcendental responsibility."[83]

Clearly, this is a natural law that Engelhardt decries, and dialogue within the parameters of this understanding of nature yields little hope for furthering moral agreements in bioethics. "One must have standards by which to judge whether what is is normatively natural or good in order to derive moral implications from nature."[84] But there is a deeper way to understand natural law that helps to explain Engelhardt's approach to the possibilities of dialogue amidst moral strangers. Natural law "properly understood" encompasses the "precepts taught us by God through our being and through the world around us, rendering a window to God."[85] Here Engelhardt leans notably on St. Isaac the Syrian who speaks of two powers of sight for man: the "bodily eyes" that gaze upon sensory objects and the "eyes of faith" that gaze by faith upon "hidden treasures." The possibilities for seeing nature rightly are found within mankind, yet this natural law within man is incomplete and "noetically opaque" apart from God. As St. Isaac states:

> The intellect is spiritual perception that is conditioned to receive the faculty of divine vision, even as the pupils of the bodily eyes in which sensible light is poured. Noetic vision is natural knowledge that is used [by power] to the natural state and is called natural light.[86]

Engelhardt is persistent that a primary issue with natural law is that nature is "ambiguous," meaning that nature can be understood in reference to God or in reference solely to humans as humans. Genesis 6:6, for example, details how the human heart is subject to "continual brooding over evil," and when the eyes of (natural) spiritual perception are dim, so, too, the eyes gazing upon the natural world are dimmed:

> The world seen through wicked eyes confirms the naturalness of natural evil and of evil inclinations. This view of reality is further strengthened as nature is resituated within

83. Goss and Vitz, "Natural Law among Moral Strangers," 287.

84. Engelhardt, *Foundations of Bioethics*, 56.

85. Engelhardt, *Foundations of Christian Bioethics*, 176.

86. Isaac the Syrian, Asc*etic Homilies*, 250–51; see Engelhardt, *Foundations of Christian Bioethics*, 173–74.

> the expectations of modern physics and biology. Everything is defined in terms of possibilities for discursive analysis, examination, study, and reasoning, but not for prayer. Construed within the truncated terms of post-Newtonian physics and post-Darwinian biology, nature becomes an all-inclusive sphere of immanence closing man off from God. As a consequence, neither physical nor biological nature is recognized as gesturing beyond itself to the Creator.[87]

In this way, Engelhardt rejects natural law theories—and other such discourses—in bioethics that posit the existence and nature of God as inessential to ethics. Natural law "properly understood," however, necessarily entails a "distinctively Christian ethic," and Goss and Vitz refer to this as the principle metaethical thesis.[88] But there is more to consider beyond solely positing the existence of God, for within the movements and transitions of how natural law is understood we can glimpse important epistemological differences. Here Goss and Vitz refer to a "principal moral-epistemological thesis of natural-law ethics" that differentiates between different natural law approaches to how one knows happiness and morality. Aquinas, for instance, argued for a natural and supernatural human *telos*—a natural and divine beatitude—with the understanding that the natural was incomplete and an "imperfect happiness" without the supernatural.[89] This twofold account of human happiness for Aquinas follows an understanding of created nature as necessarily participatory, where the natural is suspended from and participates in the supernatural to achieve its fullness. This participatory view of nature, however, changes in the later scholastic centuries (fifteenth through the seventeenth centuries) as Grotius and others effectively separate the two spheres of human happiness, creating space for natural beatitude apart from the supernatural. Moving forward, Goss and Vitz then point to a further transition in the modern period (seventeenth to eighteenth centuries) when the "theory and practices" aimed at human ends pertaining to God became clearly distinct and separated from those

87. Engelhardt, *Foundations of Christian Bioethics*, 175.

88. Goss and Vitz, "Natural Law among Moral Strangers," 288.

89. At times, Engelhardt is critical of Aquinas's "philosophical commitments" (Engelhardt, *After God*, 72, 84) that influenced the later Scholasticism of Scotus, Grotius, Ockham, and others, yet he elsewhere points to readings of Aquinas (see Bradley, *Aquinas on the Twofold Human Good*) that "separate him from the Scholastics who followed him" (Engelhardt, *Foundations of Christian Bioethics*, 220). Regardless of the connection between Aquinas and his followers, the twofold account of human happiness is a helpful distinction regarding the differences in moral epistemologies surrounding natural law.

aimed at human ends as they pertain to humans and the natural world.[90] From amidst this false dichotomy, one can then divide theology and ethics respectively as modern disciplines where the two are no longer essentially interwoven. Hence the scholastic and modern academic approach moves ever more into deflated, immanentized pursuits of philosophy as the domain of natural reason, discursive rationality, with theology likewise focusing not on prayerful illumination but perhaps on logical pursuits such as soteriology deciphered by natural theology. Within this framework, the very "*sources* and *objects* of moral epistemology" departed from traditional Christianity, paving the way for a Cartesian and Kantian grounding of ethics entirely outside of God. Summarizing these vast transitions, Goss and Vitz point to four attributes of modern, moral epistemology, particularly exemplified in the New Natural Law Theory: this theory enacts first a moral epistemology that is *distinctively irreligious*, "proceeding methodologically as if knowledge of God were not a benefit for knowledge of the moral law." Second, this theory is *nonnoetic* or nonexperiential, and hence primarily available through pure rationality and self-evident principles. This New Natural Law Theory is also *nonascetic*, "proceeding methodologically as if the seeking purity of heart were insignificant for attaining moral knowledge." Fourthly, this theory is *ecumenically egalitarian*, "in the sense that it posits what moral norms are, and ought to be, justified by public reasons that are equally accessible to any rational agent."[91]

In short, Goss and Vitz argue for an important distinction between Engelhardt's criticism and affirmation of natural law thinking. Natural law that denies the principal metaethical thesis and/or misconstrues the principal moral-epistemological thesis summarized above will inevitably lead to confusion in bioethics. Natural law rightly understood, however, offers a "window to God" through nature, and yet this encounter with nature entails a full-orbed understanding of philosophy as the love of wisdom that comes "not by learning, study, or deep analysis" on a theoretical level but through "faith, ascesis, and prayer." Furthermore, this natural law

> is not the knowing of a mere set of external moral constraints. It is a knowing along with God about how to approach God. Like the old English term "inwit" for conscience, it involves an internal knowledge that is not solitary. This knowing grows as we free ourselves from passion and turn to God through keeping His commandments. . . . In all of this, the commandments,

90. Goss and Vitz, "Natural Law among Moral Strangers," 289.

91. Goss and Vitz, "Natural Law among Moral Strangers," 290 (emphasis original).

> the law taught us by nature, are not external or empirical, but integral to bringing us to union with God.[92]

Within the *ordo* described by St. Maximus, the above type of natural knowing is "philosophical" in the qualified sense of *theoria* where *theoria* is preceded by the practiced virtues (*practike*) and then together *theoria* and *practike* end properly in theology (*theologia*). This philosophical approach employs a highly nuanced view of philosophy that is antithetical to the autonomous theorizing over nature in the scholastic mold. St. Maximus's use of "philosophy" assumes *philo* and *sophia* such that divine *Sophia* is known through ascetic virtues and nature understood through the spiritual depth of *theoria*. Necessarily, this threefold path for St. Maximus is enfolded within the rich assumptions of deification or *theosis* in Jesus Christ, and within this context, theology and philosophy aiming at *theosis* are essentially integrated spiritual practices that cannot be reproduced formulaically. St. Maximus, hence rephrases *practike*, *theoria*, and *theologia* as ascetic philosophy, natural philosophy, and theological philosophy.[93] Within this particular context, Engelhardt freely embraces discourses and "laws" on nature, and yet he is forthright in acknowledging and presupposing the nonstatic epistemology of traditional Christianity whereby "as man repents he is illumined by God and then united to God."[94]

Within the ontologically rich commitments of this premodern metaethics and moral epistemology, it may seem that public discourse—or other discursive reasoning such as within the academy—has little to no place within Engelhardt's paradigm for ethical knowing across the divides of moral strangers. After all, when created nature is defined essentially in reference to the transcendent Creator, what degree of hope remains for authentic dialogue amongst those who separate nature from the supernatural? This supposition seems logical, and yet Engelhardt clearly practices a philosophical and public discourse aimed at securing some common ground amidst moral friends and strangers. It is not difficult to see how Engelhardt's moral strangers are strange in terms of ontological, moral, and other commitments, and yet that strangeness does not necessitate enmity. Pointing out that strangeness is an honest and therapeutic maneuver, it is honest in admitting authentic difference, and it is therapeutic in seeking to map out the space between Engelhardt's premodern assumptions on theological moral meaning and the modern assumptions on morality accessible via universal, discursive rationality. In this vein, Engelhardt is not primarily Kantian or

92. Engelhardt, *Foundations of Christian Bioethics*, 176.

93. Maximus the Confessor, *Various Texts*, 161–63.

94. Engelhardt, *Foundations of Christian Bioethics*, 177.

Hegelian, even as he employs an internal critique of universal rationality via those archetypes of the modern mind. Hence it is difficult to emphasize enough the "lamentable" possibilities within Engelhardt's secular bioethics that are revealed through the game or puzzle of thin moral meaning created through agreement or consensus. Such a "foundation" via the principle of permission and beneficence is impoverished indeed.

Impoverished though it may be, however, that strangeness enables the possibility of battleground as common ground. Again, following Goss and Vitz, we can see Engelhardt not advocating for insular moral silos of silence but for a shrewd engagement across the vast ontological, epistemological, and moral differences evident in bioethics.[95] Hence Goss and Vitz identify three different conceptions of reasoning within the public square. The "*restrictive view of public reason*" follows the notion that common ground exists ideally between members of a just society on the basis of citizenship. This entails that any distinctive sectarian postures amongst citizens must be bracketed. Such a restriction may be narrowly focused on "constitutional essentials and matters of basic justice," or it may be construed in broader degrees as any "coercive activity" of the state or any social activity/morality affecting one's fellow citizens. Within such a view, the only acceptable political engagement or discourse is one that subverts "private" or "personal" beliefs in favor of purportedly "proper" public reasoning.[96] The challenge to this restrictive view exists in the difficulty and shifting nature of defining: What is the public good? Without some more fixed and transcendent starting point, the grounding behind "public good" is inevitably fluid and transitory. Hence defining the language and actions of "tolerance" that support public good versus "discrimination" remains elusive.

The second "*aspirational view of public reason*" allows for members of a just society to speak and to persuade others from their "distinctive sectarian"

95. To further this point on the constructive role of reason in Christian bioethics, Goss and Vitz point to an analogy in the Confucian approach to dialogue advocated by Ruping Fan:

"Confucian views are certainly not neutral views. . . [A neutral stance from which to start moral reasoning] does not exist. Any moral view is from somewhere and is held by somebody. Attempting to derive a substantive stance solely form pure reason is illusory because pure reason is not substantive. Reason can play a constructive role in a moral system only through close combination with certain fundamental, content-full moral assumptions and premises, from which concrete bioethical views and visions are deduced, induced, or stimulated. Such fundamental, content-full assumptions and premises can only be found in particular moral traditions, such as Confucianism. However, it is impossible to unite basic assumptions from different traditions to make a coherent moral doctrine as cosmopolitans expect. These assumptions are mutually incommensurable" (Fan, *Confucian Bioethics*, 3).

96. Goss and Vitz, "Natural Law among Moral Strangers," 293 (emphasis original).

rationalities for the benefit of their fellow citizens. In this viewpoint, motives on end-goals matter greatly even while different understandings of the means remain in contention. A physician may, for instance, advocate for his patient to avoid hastening death through medical-technological means for the good of the patient's spiritual soul, even while the patient may deny spiritual realities. Apologists within this vein "ought to strive to develop a rationalistic, universally accessible ethical theory to both ground their moral views and to guide their social-political discourse."[97]

The third "*rhetorical view of public reason*" most particularly aligns with Engelhardt's public discourse. This approach "posits no secular procedural restrictions on the ethical reasoning," nor does it encourage people "to aspire to ground their publicly relevant moral views and their social-political discourse in a rationalistic, universally accessible ethical theory."[98] Instead, this viewpoint emphasizes persuasive rhetoric as a way to cross content-full lines of sectarian difference in a prudential manner. Employing a full range of *ethos, pathos,* and *logos*, such a rhetoric aims to encourage Christian philosophers to speak boldly and without compromise of content or methodology. In short, Goss and Vitz see Engelhardt encouraging

> Christian bioethicists [to employ, as able] public reasons rhetorically to bring about Christ's ends, provided they recognize that any agreement obtained by these means does not trump the robust, traditional Christian teachings that have been revealed to the saints by noetic, liturgical ascetic encounters with the living God.

Further, the final purpose of Engelhardt's critical yet therapeutic mapping of bioethics aims to direct Christian bioethicists

> away from a kind of philosophy that, on [Engelhardt's] account, results in little more than a series of unwinnable games of puzzle play, and back toward the true philosophy, a liturgical ascetic way of life that requires a genuine love of Divine Wisdom and leads toward the fullness of human flourishing that is manifest in the lives of Christian saints.[99]

What Engelhardt's foundational approach seeks to embody is not a restrictive antagonism but prudential caution when engaging in public discourse on bioethics across lines of moral and ontological strangeness.

97. Goss and Vitz, "Natural Law among Moral Strangers," 293 (emphasis original).

98. Goss and Vitz, "Natural Law among Moral Strangers," 293–94 (emphasis original).

99. Goss and Vitz, "Natural Law among Moral Strangers," 295.

Engelhardt is not furthering a Tertullian model of "resistance and rejection" between Christian and non-Christian communities where a type of logical consistency and "manifest fervor" eliminates dialogue with anyone who does not acknowledge the supremacy and significance of Christ over all things.[100] A more fitting model is seen in Clement of Alexandria or St. Justin Martyr who allow for discerning and intelligent cooperation amidst Christians and non-Christians while maintaining the thick particularity of Christianity. What undergirds this discerning engagement, Trader notes, is a theological commitment

> that there is only one unified created reality in time and space from which the uncreated energies of the Triune God are never absent. This presence of God historically in his Church and naturally in his creation enable the fathers to make use of material gained from empirical knowledge about the workings of God's world ("natural" or general revelation through created being) and God's working in the world ("supernatural" or special revelation in history). [In short] the Church and the world form an organic unity, which means that "everything is potentially the Church and called to sanctification." Thus, when frail and moral thought opens up into the immensity of Christ the God-man, the whole panorama of creation and history are beheld from the elevated point of the mind of Christ.[101]

Such an openness moves well beyond "an isolationist Tertullian ghetto" that some may see in Engelhardt, moving instead toward an explicitly hierarchical relationship where theology is given a "place of logical priority over secular knowledge," and where Christian bioethics can engage "natural discourses" within an explicit framework of "deeper meaning and ultimate significance." The relationship is akin to envisioning the context of "a sentence governed by the rules for grammar and a group of words defined by the rules of lexicography."[102] Such a theological framework grants meaning to discerning openness over "nature" and her laws that otherwise cannot exist within the logic of the secular. This model of openness from Clement or St. Justin is a fitting analogue for Engelhardt's critical mapping, and such a context is helpful for understanding why Engelhardt stresses the critical importance of methodology for approaching morality and bioethics: "If reason is at the center of theology, one will rationally reconstruct the content of any Christian bioethics in the image and likeness of secular moral

100. Trader, *Ancient Christian Wisdom*, 13.

101. Trader, *Ancient Christian Wisdom*, 17.

102. Trader, *Ancient Christian Wisdom*, 18.

rationality."[103] Discursive reason does not accord within Christianity as a source of moral norms to be identified outside the infinite God. As such, any account of natural law "outside of a proper religious life" misconstrues theology as "faith seeking understanding rather than faith seeking holiness." Hence Engelhardt reminds us:

> The Christian *tertium quid* is the very energies of God himself. . . [and in this way] theology seeks union with God through experiencing Him, not through reasoning to Him or about Him. This is not to say that reasoning is useless or bad, but rather that it is peripheral to holiness.[104]

Despite the clarifications such as the above, Engelhardt remains a controversial figure to many who see him advocating "against philosophy" when what is needed are distinctions and proper nuances regarding what we mean by philosophy. The argument here is that Engelhardt's foundational approach to philosophy and bioethics in the public square parallels Wittgenstein's method of philosophy that seeks clarity and a change in the *form* of questions. This is a methodology that reverses the high ground where secular discursive reasoning is considered the "common ground" or neutral place from which bioethics is predominantly determined. Within such a secular framework, religious or theological bioethics becomes something aberrant, or something to be tolerated within the democratic pluralism of the twenty-first-century West. Against this secular framework, Wittgenstein and Engelhardt have left their mark through language and bioethics respectively as philosophers who question the impulse of philosophizing itself within the secular. Hence Edwards notes of the later Wittgenstein that his critiques of philosophy come from "a perspective not itself philosophical,"[105] and much the same can be said of Engelhardt. His critique is that bioethics largely functions as an extension project of the Enlightenment hope in autonomous rationality, and it is this theoretical impulse to secular philosophizing that is countered in part via Wittgenstein's general religious perspective and more robustly via Engelhardt's particular appeals to traditional Christianity.

103. Engelhardt, *Foundations of Christian Bioethics*, xv.

104. Engelhardt, "Foundations of Bioethics and Secular Humanism," 280.

105. Edwards, *Ethics without Philosophy*, 7.

6

How Sacred Cosmology Saves Ethics

The aim to this point has been to place Wittgenstein's way of thinking into conversation with the foundations of bioethics. The majority of the argument has been critical, and yet this final chapter turns to examine what Wittgensteinian thinking may offer constructively, or redemptively, to the conversation on the foundations of ethics and bioethics. This constructive aim is not systematic nor will it seek here to offer answers to particular cases in bioethics. Wittgenstein's constructive turn, rather, is relevant to the foundations of ethics and bioethics as a first-level metaethics before one may pursue particular topics and instances.

Moreover, the constructive turn in Wittgenstein is theological, and this theological turn particularly reframes ethics in an anthropological and epistemological way; importantly, this way is opened through humility as Wittgenstein's ethical approach begins with noting finite human capacities for ethical knowing. In short, where foundational approaches to bioethics such as common morality or utilitarianism appear deeply confused and embedded within modern assumptions of Cartesian personhood and theoretical ethical knowledge, this chapter offers to bioethics a profoundly different, premodern approach. This chapter seeks to change the *form* of ethics and bioethics in three ways. The first way examines Wittgenstein's religious point of view (*RW*, 94) that calls upon ethics to admit the existence of sacred and transcendent realities beyond scientific thinking and materialism. Wittgenstein may be seen in this way as one who challenges modern individuals to look beyond reductionist efficient causality to see the world as a miracle imbued with wonder.

In the second way, Wittgenstein's thinking parallels or exhibits an ancient method of apophatic epistemology that guides one toward an understanding of ethics itself as a form of apophatic and embodied knowledge. Hence Wittgenstein offers a way to change the form of our thinking, leading us toward a radically different view of ethics not as theoretical application of values—easily manipulated by man—but as ascetic formation into transcendent life and values. This change in the form of ethics is profound for all of life, both *bios* and *vios*: "we feel that even if all possible scientific questions be answered, the problems of life have still not been touched at all. Of course there is then no question left, and just this is the answer" (*T*, 6.52).

These themes lie very much within the grain of Wittgenstein's thinking, and as such these may be seen as important aspects of tracing Wittgenstein's vision of ethics and life (*Lebensanschauung*). The third and final way of changing the form of ethics takes us to sacred cosmology. On the one hand, this third step is challenging for speaking of things that are better exemplified or simply lived. Just as Wittgenstein abandoned Cambridge to teach in a rural one-room schoolhouse in the Alps, or given his persistent interests in Russian folk spirituality and monasticism, perhaps the truest expression of sacred cosmology would be to immerse oneself in a traditional Christian parish life that embraces the fullness of art, symbolism, liturgy, and other ways of life that perpetually incarnate and express in the embodied, material spheres the higher meaning or principles (*logos*) from the heavens.[1]

In a real sense, sacred cosmology is a symbolic way of speaking that draws our attention toward embodied meaning and away from the autonomous or scientific thinking Wittgenstein ardently critiqued. Pedagogically, this corrective turn from theoretical ethics to symbolic ethics is timely given the trajectory and momentum of the scientific takeover evidenced in bioethics.[2] Since the late 1960s, the literature of bioethics has raised

1. Indeed, this is part of Engelhardt's strategy—writ especially in the footnote trails of *Foundations of Bioethics*—when he points to the beauty of particular saints, of historical narratives, of parish life, and more. Admitting this higher expression, speaking of these things under language such as sacred cosmology requires thoughtful engagement; as long as sacred cosmology does not become a linguistic *techne* akin to Beauchamp and Childress's four principles, we not only *may* but *should* speak of these things carefully to communicate the truth, becoming all things to all men (I Cor 9:22).

2. The phenomenon of Jordan Peterson's rise to popularity gives testimony to the craving in modern society to move beyond the thinness of the moral self-governed by autonomous or scientific thinking: "It has become more or less evident that pure, abstract rationality, for example, ungrounded in tradition—the rationality which defined Soviet-style communism from inception to dissolution—appears absolutely unable to determine and make explicit just what it is that should guide individual and social behavior . . . . Some patterns of interpersonal interaction—which constitute the state, insofar as it exists as a model for social behavior—do not produce the ends they are supposed to produce, cannot sustain themselves over time, or even produce contrary ends, devouring those who enact them and profess their value. Perhaps this is because

foundational questions that on the one hand should be seen as variations of Anscombe's withering critique: Can modern moral philosophy maintain the moral "ought?" At the same time, these questions are distinct in that they are being taken up in a new field that has passed fifty years of activity as "bioethics." The cultural and social impact of bioethics is undeniable and does not seem to be slowing. Moreover, this concluding chapter addresses bioethics from the Wittgensteinian posture of *seeking clarity*, and recalling Kopelman's analysis of bioethics as a second-order discipline,[3] Wittgenstein's ethical vision serves to remind bioethics of the deeper and mysterious nature of ethics. In this vein, the argument here is that for bioethics to function authentically, the metaethical questions that are inherently anthropological, epistemological, and religious must be engaged. Bioethics should not seek to narrow its scope or methodology within one of its existing professions or fields but rather to expand beyond them. Within a theological mode of bioethics, this entails envisioning ethics as a process for human formation within end goals and ideas on the good. In this way, Wittgenstein's ethical vision explored in this last chapter may be seen as a Socratic dialogue relevant to bioethics. As the quest to understand virtue comes to life through dialogue with Gorgias's student, Meno, so Wittgenstein may be taken as a pedagogue amidst the perplexities (*aporia*) and puzzles (*narke*) at the foundations of bioethics.

Beyond Scientific Ethics: Wittgenstein's Religious Way of Thinking

> *I am not a religious man but I cannot help seeing every problem from a religious point of view.* (*RW*, 94)
>
> *Believing means submitting to an authority. Having once submitted, you can't then, without rebelling against it, first call it into question and then once again find it acceptable.* (*CV*, 45e)

Here at this point, we can trace Wittgenstein's constructive vision that moves past the trappings of scientific thinking, and for Wittgenstein this

planned, logical and intelligible systems fail to make allowance for the irrational, transcendent, incomprehensible and often ridiculous aspect of human character. . ." (Peterson, *Maps of Meaning*, 22).

3. "*A second-order discipline* has members who (1) generally do not seek to make their subject matter narrower; (2) envision the problems they address as complex and profiting from the views of people of different professions, disciplines, and academic fields, and seek to broaden the perspective and expertise available to address those problems; and (3) rely in part upon the primary professions, disciplines, or academic fields to set their own relational or other standards of competency for their members" (Kopelman, "Growth of Bioethics," 152 (emphasis original).

move interweaves nontheoretical ethics with religious (*religare/religio*) thinking.[4] This is not to say that Wittgenstein is an effective guide for seeking a particular, richly dogmatic theological ethics. Rather, the figure of Wittgenstein—his thinking and lived life—presents a pedagogical example well suited to the discussions surrounding the foundations of ethics and bioethics. Wittgenstein remains a religious generalist who presents puzzles and practices of clarity to deflate scientific thinking. For whatever the biographical reasons we may know or suspect, Wittgenstein seemed unable to move beyond his prescribed practices of clarity such that even while admitting that the framework or end of his philosophical practices is the mystical (*T*, 6.522, 6.53), he did not seem to apply the hints of the transcendent within his writings to find some form of personal, spiritual conclusions.[5] Up to this last chapter, Wittgenstein's critique has been prominent, and yet the whole of that critique belongs within "the fundamental intention of Wittgenstein's thinking . . . to incarnate a vision of the healthy human life; the transmission of a moral vision."[6] Moreover, in a real sense, this moral and religious vision is a poetic and ascetic endeavor as Wittgenstein seeks to move past re-presenting the world in theoretical language but rather to

4. *Religare/religio* in Latin retains the meaning of "bond" between the sacred and mundane, or "reverence" for the gods; religious thinking draws together extremes, uniting disparate things like ligaments (*ligare/ligamentum*) in the body.

5. Kallenberg reminds us of the importance in considering Wittgenstein's lived life as a way to read his writings in the right spirit, and Kallenberg especially points to Wittgenstein's "hermetic existence, the giving away of his share of the family fortune, his serious consideration of becoming a priest or monk, his retreat from academics to teach schoolchildren in Austria" all as features that help explain long-standing attraction to the Christian gospel as he encountered it through Tolstoy's *Gospel in Brief*. Much, much more stands to be considered in theological terms of what might constitute communion with the mystical, but for Wittgenstein it is clear that much of his life—at least since 1914—was lived searching (Kallenberg, *Ethics as Grammar*, 172). Consider also Wittgenstein's own words in *Culture and Values*: "What inclines even me to believe in Christ's Resurrection? It is as though I play with the thought—If he did not rise from the dead, then he decomposed in the grave like any other man. *He is dead and decomposed.* In that case he is a teacher like any other and can no longer help; and once more we are orphaned and alone. So we have to content ourselves with wisdom and speculation. We are in a sort of hell where we can do nothing but dream, roofed in, as it were, and cut off from heaven. But if I am to be REALLY saved—what I need is *certainty*—not wisdom, dreams or speculation—and this certainty is faith. And faith is faith in what is needed by my *heart*, my *soul*, not my speculative intelligence. For it is my soul with its passions, as it were, with its flesh and blood, that has to be saved, not my abstract mind" (*CV*, 33, emphasis original). Likewise, Monk's biography offers many helpful insights into Wittgenstein's influences and fascinations with music, poetry, duty, sexuality, Catholicism, monasticism, etc. (Monk, *Ludwig Wittgenstein*, 3–27, 551–75).

6. Edwards, *Ethics without Philosophy*, 203.

present the world and to live in it. Edwards draws attention to this Heideggerian connection in Wittgenstein:

> The world—manifold, indefinitely various, self-concealing, and self-revealing—is there in the poems itself, but not literally in the images that constitute them. The poet wants to *present* the world not re-present it. He wants to get us to see ourselves in all our seeing, to break the grip of the self-forgetfulness of representation. The poet's images are not statements; they are reminders, reminding us, putting us in touch with all the images through which we see, deliteralizing our perceptions.[7]

This poetic dimension in Wittgenstein further reinforces the need to understand the whole of his life and work. Wittgenstein noted that "the philosophical remarks in [the *Philosophical Investigations*] are, as it were, a number of sketches of landscapes which were made in the course of these long and meandering journeys" (*PI*, 3e). His work is an "*ethical deed*" and an aid both to "thinking and living,"[8] and the point that draws all of this human understanding together is Wittgenstein's regard for *das Mystiche* and the essentially *religious* thinking that is needed to engage sublime truth beyond the scientific. For Wittgenstein, scientific thinking within limits is hardly problematic, and as one who studied and valued engineering and architecture, we should not take a simplistic view of Wittgenstein. As with transscience, however, the problem arises when scientific thinking progresses beyond its natural sphere and misses the proper religious mode of thinking for truths that transcend quantitative observations. Hence in *PI*, 110, it is ironic that Wittgenstein points out that "superstitions" are found in metaphysical philosophers who seek theoretical truths out of the forms of language, assuming a "depth" of meaning where it is not present. "The problems arising through a misinterpretation of our forms of language have the character of *depth*" (*PI*, 111, emphasis original), and in this way superstition arises not in ritual, religious practice but rather in metaphysical philosophy. As Edwards notes, ritual sees and appreciates a broad and deep *Pathos* (impressiveness and formation) within liturgical or traditional actions, whereas the metaphysical philosopher thinks reductively by claiming ritual as mere

7. Edwards, *Ethics without Philosophy*, 214 (emphasis original), 203–16. Edwards continues to note that the later Wittgenstein's "philosophical therapy is not exactly poetry" but rather that "both flow from the same deliteralizing sensibility. The metaphysical philosopher is susceptible to being captured by grammatical pictures interpreted as literal representations of the superfacts underlying language. To be freed from such captivity it is necessary to be freed from the literalizing sensibility that promotes it" (214).

8. Edwards, *Ethics without Philosophy*, 214 (emphasis original).

superstition. Hence in the presence of *das Mystiche*, the scientific philosopher's pathological superstition is to seek an explanation by drawing on a mental reservoir of assumed meaning (*BB*, 143) whereas the ritual commoner responds with deep *mythos*.[9]

For Wittgenstein, it is pathological to create explanations via philosophical language to explain the world before us. "It is not *how* things are in the world that is mystical (*das Mystiche*), but *that* it exists" (*T*, 6.44, emphasis original), and this religious point is intimately connected to ethics. The scientist views the world as a series of "facts," whereas the ethical view of the world is to view it as a miracle:

> And there, in my case [when seeking absolute or ethical value] it always happens that the idea of one particular experience presents itself to me which therefore is, in a sense, my experience *par excellence*. . . . I believe the best way of describing it is to say that when I have it *I wonder at the existence of the world*. And I am then inclined to use such phrases as "how extraordinary that anything should exist" or "how extraordinary that the world should exist." (*LE*, 10–11, emphasis original)

> . . . it is absurd to say "Science has proved that there are no miracles." The truth is that the scientific way of looking at a fact is not the way to look at it as a miracle. (*LE*, 16–17)

Calling attention to science, miracles, and ritual, Wittgenstein seeks to turn the form of our thinking from philosophical propositional thinking to a type of religious thinking.[10] Here it is helpful to consider Norman

9. Even the tossing of spilled salt is an occasion to note the difference between the scientific materialist who cries "superstition!" and the ritualist who sees *pathos*. "Perhaps in this case the *Pathos* is connected to our awareness of salt as savor and preserver: as easily as we carelessly let spill these valuable crystals, so easily can we let spill the things that give our life savor and that preserve us from spoilage; and the throwing ritual is a way of acknowledging our carelessness and our determination to do better. But whatever the source of the *Pathos*, the important consideration is that in the ritual is a response to some deep feeling" (Edwards, *Ethics without Philosophy*, 232).

10. Drawing attention to the religious dimensions in Wittgenstein is not a new suggestion, and yet the goal in tracing this religious thinking is to examine a pattern of thinking that is relevant for the foundations of ethics and bioethics. "It is almost platitudinous to state that there is a profound religious dimension to the work of Ludwig Wittgenstein. Nevertheless, this attribution of a religious quality may seem bizarre when one is confronted with his two masterpieces—*Tractatus Logico-Philosophicus* and *Philosophical Investigations*—which seem to have little application to life, let alone to affairs of the spirit. Yet Wittgenstein's work became associated with religion when sympathetic philosophers applied his notions of 'language-games' and 'forms of life' to religious belief and practice. Closer attention to Wittgenstein's philosophy (and how he perceived it) then revealed that much of his writing was conditioned by religious

Malcolm's four analogies between Wittgenstein's philosophy and religious thinking. After reviewing Malcolm's analogies below, we will then probe further into Malcolm's first analogy by contrasting "intellectualism" with religious thinking, and in the third point I draw attention to the way that Wittgenstein's thinking does not inaugurate a theoretical philosophy of religion but rather supports an important premodern view of reasoning in relation to authority, mysticism, and more.

For Wittgenstein's follower Malcolm, Wittgenstein's philosophical outlook is *analogous* to religious thinking. Malcolm does not see a "resemblance" between the two, but rather sees analogies particularly, the first of which relates to "explanation," or how explanation "reaches a limit, and when pressed further loses its sense."[11] The term explanation here is key for the way that it relates to Wittgenstein's critique of philosophy as a seemingly limitless process of scientific theorizing. Wittgenstein reacted strongly against the idea that philosophy is capable of explaining *everything* and believed that in place of endless "rational justification" philosophy should be seen within the context of *gift*.[12] As God grants the gift of faith—a gift yielding its own rationality or internal logic, so to speak—the expectations for rationality shift from a discourse of physical and material causality to the language of acknowledging God's will, whatever that may be.[13] Acknowledging that "God wills it" is not to claim that such is a sufficient, final answer, say, to Job's suffering; such words are not uttered as an "explanation" for Job's suffering but rather as a functioning language to address some profound, unexplainable state of reality, and within a Wittgensteinian way this reality may be called the nature or form of the matter (*PI*, 654). Malcolm notes: "The function of the words, 'It is God's will,' when said religiously and seriously, in time of trouble, is not to offer the *final* explanation at all. Instead, they are an attempt to bring to an

considerations" (Clack, *Wittgenstein, Frazer and Religion*, 3). Further, others move beyond Wittgenstein's allusions and passing comments on Christianity in particular to draw strong parallels and conclusions between Wittgenstein's thought and Christian theological discourses proper (see Barrett, *Wittgenstein on Ethics and Religious Belief*, and Labron, *Wittgenstein and Theology*). General analogical thinking may be helpful, but theological particularity certainly does not stem from Wittgenstein himself.

11. Malcolm, *Wittgenstein*, 84.

12. Malcolm, *Wittgenstein*, 84.

13. Parallels to types of natural law thinking are appropriate here; deistic or even atheistic "new natural law theory," for instance, that distances (or brackets entirely) natural law thinking from the divine law-giver emphasizes patterns within nature without asking the more fundamental question: What is the source of nature? Who is the divine intelligence or creator of the natural world that emulates in times and seasons certain laws or patterns?

end the torment of asking 'Why did it have to happen?'—an attempt to give the tormented one rest, to provide *peace*."[14]

Malcolm's aim with the example of suffering is to illustrate the fundamental need "to accept" the way many things are. This is not a call to omit critical thinking and reflection on mysterious realities,[15] but rather to note the mystery that pervades reality and alters our mode and approach for knowing reality. Malcolm's point draws attention to the theme of "description" over "explanation" throughout the *Philosophical Investigations*:

> Wittgenstein regarded the language-games, and their associated forms of life, as beyond explanation. The inescapable logic of this conception is that the terms 'explanation,' 'reason,' 'justification,' have a use *exclusively within* the various language-games. The word 'explanation' appears in many language-games, and is used differently in different games. My explanation of *my* motive, for example, is a different conception of 'explanation' from my explanation of *your* motive. An explanation of why your car won't start will be radically different in kind from an explanation of why a friend of yours is avoiding you.
>
> An explanation is *internal* to a particular language-game. There is no explanation that *rises above* our language games, and explains *them*. This would be a *super-concept* of explanation—which means that it is an ill-conceived fantasy."[16]

In this way Wittgenstein's language games are analogous to old-world religious thinking that humbly accepts facts or reality as the "way things are," and this attitude carries implications not only for the limits of philosophy but also for natural science and medicine. For instance, are we "explaining" reality when stating that gravity is the force that draws bodies to fall toward the earth, or are we—in certain words within a concrete language game—describing one way of addressing this reality? Are we explaining the reality of the body and cancer when discovering that a certain chemotherapy slows down the spread of the cancer? Likewise, Malcolm

14. Malcolm, *Wittgenstein*, 86 (emphasis original).

15. Though not necessarily an attempt to shortcut critical thinking where it is due, I will discuss further below the tendency in Malcolm's first analogy to promote a validation of belief *because* philosophical explanation is inherently flawed, or to assert that "religion is a form of life" that disallows those outside the form of life to carry on rational discussions with those inside, say, Christian religion. This latter "Wittgensteinian fideism" is problematic if for no other reason than it often serves as a caricature and misses the point of Wittgenstein's *Lebensformen*. See Kerr, *Theology after Wittgenstein*, 28–31; Incandela, "Augustinian Reading of Aquinas and Wittgenstein," 33–35.

16. Malcolm, *Wittgenstein*, 77–78 (emphasis original).

questions what we can explain *essentially* about the existence of sports, or of musical composition.[17] "Our mistake is to look for an explanation where we ought to regard the facts as 'proto-phenomena.' That is, where we ought to say: *this is the language game that is being played*" (*PI*, 654, emphasis original). Viewed in this way, Wittgenstein's language games are less susceptible to charges of asserting reality that is ontologically "created" by language games—a sort of philosophical metaphysics from grammar—for the simple reason that language games are a description, and description is a polyphonic way of engaging reality. A scientific language game is hardly "flawed" for its particular attention to natural phenomena but rather for its tendency in modernity to claim exclusive explanations of the essence of a thing supposedly hidden somewhere "behind" or "within" the thing (*PI*, 435–43). "The truth is that the scientific way of looking at a fact is not the way to look at it as a miracle" (*LE*, 17). Likewise, logic is problematic when the "hardness of the logical must" trumps all other modes of apprehending reality (*PI*, 437). Language games are not an exhaustive philosophical structure or metaphysical grounding but function more in a critical way for understanding some of the ways things are as they simply unfold in human ways of language and descriptions of reality.

Building upon Malcolm's analogy, we may align Wittgenstein's language games as religious thinking with patterns of Patristic economic language in seeking to explain with the incarnation. Without anticipating too much the discussion to follow below on apophasis, here we do well to remember the distinction between theology proper (*theologia*) that seeks union with God, particularly in asceticism, prayer, and liturgy, and economic theology that contemplates and speculates on the scriptural and other descriptions of Christ. The sufferings of the divine word in Christ are rightly spoken of according to the economy (*kat' oikonomian*) whereas such language would be blasphemous "according to the deity" (*kata theoteten*).[18] To be sure, the reference point for Wittgenstein's language games is far from the particular Trinitarian language of patristic theological discourse, and yet Malcolm's first analogy draws attention to the way Wittgenstein's clarifying language games escape philosophical metaphysics by adhering to a religious mystery that might truly ground philosophical thinking.[19] All

17. Malcolm, *Wittgenstein*, 85.

18. McGuckin, *Westminster Handbook to Patristic Theology*, 112.

19. Wittgenstein's use of the term "metaphysics" carries a technical meaning: "'metaphysical' indicates an invalid move within a language game—an attempt to express what can either not be expressed at all, or not in the given way. Such futile efforts, according to Wittgenstein, spring from misunderstanding the grammar of our language. The dualist construal of sensation as a mental object is the product of misunderstanding

language, contemplative descriptions, and other claims exist in a suspended relationship according to the economy of this mystery. Wittgenstein never names the mystery with dogmatic particularity, but the important thing to note is the pattern of his thought that pushes beyond the secular to the religious, allowing for a type of authentic philosophical thinking and language as description rather than explanation of *das Mystiche*.

Malcolm's second analogy furthers the first by focusing on the sense of "wonder" that properly arises from "seeing the world as a miracle" (*LE*, 17). Beyond theoretical, scientific speculations regarding "the first state" of the universe, Wittgenstein recognized a deeper layer of questions that science cannot address: namely, *why anything exists at all*.[20] The only proper stance to those questions of deepest, theological meaning for Wittgenstein is wonder. "Man has to awaken to wonder—and so perhaps do peoples. Science is a way of sending him to sleep again" (*CV*, 5). Moreover, Malcolm does not hesitate to point out the relationship between that profound, cosmological wonder and the wonder of certain language games:

> In the *Investigations* and other late writings, Wittgenstein sometimes expressed a kind of wonder at the existence of the various language-games and their contained forms of human action and reaction. 'Let yourself be struck by the existence of such a thing as our language game of: confessing the motive of my action' (PI, p. 224).[21]

It is important to note that Malcolm's point on Wittgenstein's wonder does not mean that language games themselves are in some way sacred, but rather that the religious sense of seeing the world as a miracle and a gift opens the way for philosophical conceptions that embrace simple and profound astonishment. In a poetic way, Wittgenstein states: "Now I am tempted to say that the right expression in language for the miracle of the existence of the world, though it is not any proposition in language, is the existence of language itself" (*LE*, 17).[22] The point is to accept a feeling of

---

the grammar of the mental and an example of such (bad) 'metaphysics'" (Sidiropoulou, "Wittgenstein's Apophatic Descriptions," 58–59n11).

20. Malcolm, *Wittgenstein*, 87.

21. Malcolm, *Wittgenstein*, 87.

22. On a related note, Wittgenstein saw philosophy done well as a "poetic composition": "I think I summed up my attitude to philosophy when I said: philosophy ought really to be written only as a *poetic composition* [*Philosophie dürfte man eigentlich nur DICHTEN*]. It must, as it seems to me, be possible to gather from this how far my thinking belongs to the present, future or past. For I was thereby also revealing myself as someone who cannot quite do what he would like to be able to do (*CV*, 24, emphasis original).

wonder and religious sense of miracles within philosophical activities. The practices of language owe much to wonder as a way to move beyond communication as a bare mechanism:

> I'd like to say: you regard it much too much as a matter of course that one can communicate anything to anyone. That is to say, we are so much accustomed to communicating in speech, in conversation, that it looks to us as if the whole point of communicating lay in this: that someone else grasps the sense of my words—which is something mental—that he, as it were, takes it into his own mind. If he then does something further with it as well, that is no part of the immediate purpose of language. (*PI*, 363)

The third analogy is one Malcolm nuances so as not to overstate. It is an analogy between the sickness of the spirit—a religious concern—and the intellectual diseases of theoretical philosophy. The commonality in both is a view that "something is wrong with us," and this view of an "illness" and "disease" (*Krankheit*) informs the activities of religion and philosophy for Wittgenstein.[23] Deep within the practice, thinking, and emotions of religion must be some recognition of human imperfection and need for healing from some ancestral sickness. Within philosophy, a similar motivation prompts a search for explanations, and the process goes awry when we search for explanations from a sort of metaphysical "reservoir" from which flow reasons and rationale. Malcolm notes that:

> this same movement of philosophical thinking is sometimes described by Wittgenstein as the postulating of 'intermediary steps.'
>
> We are treating here of cases in which, as one might roughly put it, the grammar of a word seems to suggest the 'necessity' of a certain intermediary step, although in fact the word is used in cases in which there is no such intermediary step. Thus we are inclined to say: 'A person *must* understand an order before he obeys it,' 'He must know where his pain is before he can point to it,' 'He must know the tune before he can sing it,' and suchlike. (BB, p. 130)
>
> The craziness of much of philosophical theorizing comes from yielding to the temptation to explain everyday actions, reactions, abilities, by inventing 'reservoirs' of mental states, intermediary steps, underlying mechanisms.[24]

23. Malcolm, *Wittgenstein*, 87–90.

24. Malcolm, *Wittgenstein*, 88.

For Wittgenstein, a philosopher must fight hard against seeking to invent "hidden states" or "processes" to explain what only may be described. "The philosopher treats a question; like an illness" (*PI*, 255), and "Our illness is this, to want to explain" (RF, 333). Therapeutic philosophy for Wittgenstein is an activity that begins with a desire for healing, and this is contrasted with the ambition to uncover a nonexistent, autonomous metaphysical reservoir. Therapeutic philosophy exists with a formative, ascetic purpose aimed at a final good, and such a philosophy blurs the lines with theological thinking.

The fourth analogy addresses what we might call Wittgenstein's concern with natural theology. Wittgenstein saw little value in intellectual proofs for God's existence, and this led him typically to emphasize the role of acting and doing rather than reasoning and interpreting. Hence Malcolm supposes that Wittgenstein would very much have favored St. James's particular message that "Faith, without works, is dead" (Jas 2:17).[25] When speaking once with Maurice O'Connor Drury on the latter's intentions to seek ordination, Wittgenstein cautioned Drury against a way of thinking in many parsons that—along with Bertrand Russell—had "done infinite harm, infinite harm." Wittgenstein explained:

> I would be afraid that [Drury] would try and give some sort of philosophical justification for Christian beliefs, as if some sort of proof was needed. You have intelligence; it is not the best thing about you, but it is something you mustn't ignore.—The symbolisms of Catholicism are wonderful beyond words. But any attempt to make it into a philosophical system is offensive. (*RW*, 117)

In that same conversation, Wittgenstein urged Drury not to forget the existence of the Eastern Orthodox Church and remarked that there had only been "two great writers in Europe of recent times, Tolstoy and Dostoevsky" (*RW*, 117). What is common amongst these strands of thought is Wittgenstein's aim to see actions and deeds existing primarily "at the base of our concept of human being, of a being with mind and soul." Continuing on with the point, Malcolm notes Wittgenstein's persistent attempt "to locate the basis of our concepts in prelinguistic, prerational actions and reactions. It is not from intuitions, nor convictions, nor any kind of reasoning, that our language-games emerge—but from 'our acting'" (*OC*, 204).[26] In short, Malcolm's analogies draw out a vision of Wittgenstein as a philosopher of

25. Malcolm, *Wittgenstein*, 90.

26. Malcolm, *Wittgenstein*, 92.

meaningful, ascetic, or formative embodiment who employs a critique of autonomous theorizing where religious thinking is more fitting.

Moreover, here it is helpful to return to Malcolm's first analogy as a way to contrast "intellectualism" with religious thinking, especially in reference to Wittgenstein's "Remarks on Frazier's *Golden Bough*." In this work, Wittgenstein does not hide his dissatisfaction in secular theorizing about religious modes of knowing. To reduce the depth of religious rituals and myths into explanatory theories is to miss the depth of religious thinking. Frazer's secular presupposition is that religious and magical views of mankind share a bald commonality: both are erroneous before the primacy of science and naturalism (RF, 119). Wittgenstein challenges Frazer's secular sociology:

> Was Augustine in error, then, when he called on God on every page of the *Confessions*? But—one might say—if he was not in error, surely the Buddhist holy man was—or anyone else—whose religion gives expression to completely different views. But *none* of them was in error, except when he set forth a theory. (RF, 119, emphasis original)

In a real sense, Wittgenstein is operating within a religious mode that runs directly contrary to religious studies, at least religious studies understood as the normativizing and theorizing of religious rituals from the vantage point of the secular academy. To seek out the "aims" and rationale for rituals is to miss the nature of what we may call Wittgenstein's anthropology of *homo ritualis* (RF, 129).[27]

> Burning in effigy. Kissing the picture of one's beloved. That is obviously not based on the belief that it will have some specific effect on the object which the picture represents. It aims at satisfaction and achieves it. Or rather: it *aims* at nothing at all; we just behave this way and then we feel satisfied. (RF, 123, emphasis original)

Wittgenstein laments Frazer's simpleton thought that savages did not think technically and scientifically when facing technical needs: "The same savage, who stabs the picture of his enemy apparently in order to kill him, really builds his hut out of wood and carves his arrow skillfully and not in effigy" (RF, 125). Wittgenstein's aim here is to change the form of how Frazer and other theorists see scientific and technological thinking within a dualistic schema of (mechanistic) "reality" vs. magical, superstitious *mythos*. One of Frazer's chief downfalls is not the lack of a spiritual

27. See Kerr, *Theology after Wittgenstein*, 69–76, 156–60.

life but rather the "narrowness" of his spiritual life that moves in accord with the self-proclaimed superiority of Victorian cultural secularism. "As a result: how impossible it was for him to conceive of a life different from that of the England of his time! Frazer cannot imagine a priest who is not basically a present-day English parson with the same stupidity and dullness" (RF, 125). The great irony is that Wittgenstein sees Frazer's clinical intellectualism as the more pernicious savagery:

> Frazer is much more a savage than most of his savages, for they are not as far removed from the understanding of a spiritual matter as a twentieth-century Englishman. *His* explanations of primitive practices are much cruder than the meaning of these practices themselves (RF, 131).

Wittgenstein further notes that Frazer's explanations are but one possible way of assembling the reality of the matter, and Frazer is limited by subjugating all other transcendent myths and contexts under the temporal hypothesis of his one scientific theory that inevitably hinges on some "secret law" (RF, 133). Hence Frazer's condescending assessment of religious rituals and magic is not that they are entirely devoid of rationality, but rather that they exhibit a childish rationality that has yet to mature. As Clack describes it, "the central tenet of Frazer's account is clear: both magic and religion are elementary and erroneous theories of the nature of the universe, pathfinders for science."[28] With this presupposition in mind, the essence of Frazer's religion becomes its theoretical foundation—whatever theory that may be—and all rituals become practices to be quantified and measured according to the presumed theory at hand.

Here we can discern a connection between rituals and ethics for within the secular paradigm both fall prey to reductivism under the mantle of "theoretical application." The progressions from theory to ritual or from theory to ethics both draw upon a Cartesian tradition of "mentalist-individualist" anthropology.[29] In contrast, as Kerr points out, Wittgenstein draws attention to the animality rather than to *res cogitans* of human beings, and as an example of this, Kerr suggests that this embodied focus carries implications for the status of embryos for research and abortion. Whereas Peter Singer challenges the rights of human embryos—lacking full mental capacities, perhaps consciousness—above other conscious animals on the grounds that we lack evidence that embryos experience pain,

28. Clack, *Wittgenstein, Frazer and Religion*, 11.

29. Kerr, *Theology after Wittgenstein*, 174.

Kerr proposes a Wittgensteinian consideration where *Lebensformen* as the natural a priori for human existence alters the form of the debate:

> The embryo already exists in the closest relationship of physical dependence on an adult member of our species. Our being human has to be redescribed as our being animals of a certain kind, sharing from the beginning certain possibilities of interaction and response at the very physical level of vital functions. Paradoxically enough, the more animal we remember ourselves to be, the weightier the theological objections to abortion and embryo experimentation might become.[30]

As the form of the question changes, however, we are not creating some new Wittgensteinian-approved *explanation*. For instance, drawing out features of Wittgenstein's attitude toward explanation, Malcolm states that "religious practice itself is a language-game,"[31] and we do well, following Incandela, to note the tendency in Malcolm and others to hold up one element of Wittgenstein's thinking—in this instance, language games—as an ontological grounding of some sort.[32] Incandela compares this misapplication of Wittgenstein to readings of Aquinas's Five Ways as "proofs" or reasons for God "before one believes in God." The difference here lies between two viewpoints of Aquinas's Five Ways: in the first, the Five Ways are a "requirement" of discursive reasoning that is antecedent to faith; in the second, the Five Ways are a postcedent practice of "retrospective justification."[33] For Incandela, much is at stake here, for if one reads Aquinas in the former way, Aquinas becomes a type of philosopher relying upon formal arguments "apart from a distinctively Christian *telos*, rather than a theologian who appealed to the authority of God, and hence the gratuity of God in

30. Kerr, *Theology after Wittgenstein*, 176–77.

31. Malcolm, *Wittgenstein*, 85.

32. As an example of how some read Wittgenstein's *Lebensformen* for ontological grounding, Saidel compares Agamben's understanding of *Lebensformen* to that of Wittgenstein. "Agamben reads [Wittgenstein on the difficulties of speaking of ethics or wonder in theoretic language], maintaining that what Wittgenstein calls 'the mystical' is not a psychic reality beyond language, but *infancy*: the only possible non-linguistic (and so pre-subjective) experience for human beings as the transcendental condition of the acquisition of language. What Agamben is trying to do is to turn what has been thought in Cacciari's reading of Wittgenstein as a negative foundation (the mystical, *sygetics*) as a positive condition of pure potentiality he names *infancy*" (Saidel, "Form(s)-of-Life," 172, 163–86, emphasis original).

33. Incandela, "Augustinian Reading of Aquinas and Wittgenstein," 32. In articulating this point, Incandela points especially to chapter 4 of Preller, *Divine Science and the Science of God*.

revealing and redeeming, to bring all things under and to Christ."[34] In a comparable way, language games may be misapplied to justify religion or to deconstruct philosophy "to make room" for faith. Contesting this shallow reading, Incandela and Clack, amongst others, have argued against extending language games into a philosophy of religion when Wittgenstein surely would have contested such a systematic construction.[35] Rather than reading Wittgenstein systematically—risking the frictionless pseudoobjectivity he frequently contested (*PI*, 107)—we ought to remember Wittgenstein's call for the philosopher to assemble "reminders for a particular purpose" (*PI*, 127). In this spirit, we can see that much of Wittgenstein's purpose is aimed at therapeutic reminders in this scientific age of the impossibility of "ideal exactitude given to us *a priori*" for "at different times we have different ideals of exactitude; and none of them is supreme" (*CV*, 37e). In place of such exactitude, Wittgenstein points to finitude, contingency, and the end of justification (*PI*, 217) as essential qualities that apply to scientific, philosophical, and theological modes. Hence Incandela notes a problem:

> [T]hose who use Wittgenstein for some wholesale justification of religious belief have not followed him in abandoning that project. Nor, despite appearances, have the attempts to insulate religious belief from criticism by hermeneutically sealing it into separate language-games or forms of life. The quest for universal intelligibility does not disappear with such attempts. It just gets franchised out, with each franchise sharing the values of the parent company, but now imbued with its own local flavor and insisting on its own autonomy.[36]

34. Incandela, "Augustinian Reading of Aquinas and Wittgenstein," 32.

35. See Clack, *Introduction to Wittgenstein's Philosophy of Religion*, 78–105, and Incandela, "Augustinian Reading of Aquinas and Wittgenstein." Incandela summarizes: "When Aquinas and Wittgenstein are used to resolve various epistemological crises, they inevitably get misread. It's like doing a spell check on a text of a play. The point lies not in a scrutiny of a sort that its authors would regard as peculiar. The point lies in the performance" (44–45).

36. Incandela, "Augustinian Reading of Aquinas and Wittgenstein," 35 (emphasis original). Furthering this passage, Incandela then quotes Clack: "Far from defending religion from attack, the Wittgensteinian is stripping it of any substantial content. Viewed in this manner, Wittgensteinian philosophy of religion is not a protective strategy but a full capitulation to positivism. Accepting that after Hume and Ayer there can be no way of justifying the metaphysical claims made by religion, Phillips and others of his ilk choose, from some kind of nostalgic yearning, to preserve the language of religion while rejecting the objects to which that *language* had formerly been believed to refer" (Clack, *Introduction to Wittgenstein's Philosophy of Religion*, 101).

As some seek to justify religion after Wittgenstein's critique of language, they almost surely remain grounded in the mode of critique that Wittgenstein himself saw as a practice of clarity, not a final content-creating proposal. For this reason, Wittgenstein contested the notion of a "school" or system of thought that would follow in his steps (*CV*, 61e). A Wittgensteinian school or system that would ground a system of religion or any other system misses the point; Wittgenstein is more concerned with the nearness or immediacy of practices of life that surpass theoretical systems in the way that the narrative of a story or the reading of a poem surpasses literary theorizing. As Incandela notes, aligning Wittgenstein's lived philosophical practices with narrative is precisely the way Hauerwas emphasizes the "grammar of language *used* by believers" amidst the formative thickness of community rather than the thinness of "some general account of 'human experience.'"[37] Pace Malcolm, the point is not to see religion *as* a language game but rather to see the practices of language *directing* one to the more robust reality of life itself beyond theories (or the individual pages, paragraphs, spelling in a book). In this way we are correct to note that language games for Wittgenstein do not support a religious system as much as they direct one toward a religious view of *life itself*. Hence the point Incandela makes so well is that we misread Wittgenstein to think of language games epistemologically rather than morally. To conceive of language games as an epistemological structure, one may then see language games as a way of protecting and guaranteeing some internal consistency for small communities. Questions must be left at the door, for the sense of things is internal. Viewing language games in a moral way, however, we are better able to understand this quote from Wittgenstein:

> Christianity is not based on a historical truth; rather, it offers us a (historical) narrative and says: now believe! But not, believe this narrative with the belief appropriate to a historical narrative, rather: believe, through thick and thin, which you can do only as the result of a life. *Here you have a narrative, don't take the same attitude to it as you take to other historical narratives!* Make a *quite different* place in your life for it.—There is nothing *paradoxical* about that! (*CV*, 32e, emphasis original)

Wittgenstein speaks here of religious faith that deeply values authority, a critical element of life that is lost on secular reasoning. For Wittgenstein, no doubt thinking in accord with Kierkegaard, this authority in religious thinking takes the form of trust over fear. "Religious faith and superstition

37. As quoted in Incandela, "Augustinian Reading of Aquinas and Wittgenstein," 39–40 (emphasis added).

are quite different. One of them results from *fear* and is a sort of false science. The other is trusting" (*CV*, 73e, emphasis original). Incandela summarizes:

> This is why it must be a misinterpretation of Wittgenstein to put religion forward as a self-enclosed language-game or form of life. For he is elsewhere clear that the trust about which he speaks is a 'passionate commitment' (CV, 64e). As Dallas High reminds us, passions tend to be absorbing. They get in the way of other things, and hence are not generally the kinds of things susceptible to tidy enclosure. High also appropriately draws attention to how thoroughly Kierkegaardian this is. That is appropriate because Kierkegaard, an Augustinian through his Lutheranism, was considered by Wittgenstein 'by far the most profound thinker of the last century.' And certainly, submission to authority is at the heart of Kierkegaard's discussion of Abraham. The teleological suspension of the ethical and the paradoxes of Abraham's faith require an exalted notion of authority, or else they are not in the least bit compelling.[38]

From this point, it is a short move to then see the importance and necessity of tradition in Wittgenstein's premodern thinking: "Tradition is not something a man can learn; not a thread he can pick up when he feels like it; any more than a man can choose his own ancestors. Someone lacking a tradition who would like to have one is like a man unhappily in love" (*CV*, 76e). Talal Asad elaborates on this point:

> For Wittgenstein, in other words, tradition represents for someone who doesn't have it the object of an unattainable longing: the condition of belonging to another, being accepted as such by him or her, and of hoping to learn (and construct) through friendship who one is. Of course the language and practice of tradition can and must be learnt (people do enter traditions they have not inherited) but Wittgenstein's emphasis is on the fact that what is learnt is not a doctrine (rules) but a mode of being, not a thread one can pick up or drop whenever

38. Incandela, "Augustinian Reading of Aquinas and Wittgenstein," 40. Further, Wittgenstein held that: "In religion every level of devoutness must have its appropriate form of expression which has no sense at a lower level. This doctrine, which means something at a higher level, is null and void for someone who is still at the lower level; he *can* only understand it *wrongly* and so these words are *not* valid for such a person. For instance, at my level the Pauline doctrine of predestination is ugly nonsense, irreligiousness. Hence it is not suitable for me, since the only use of it I could make of the picture I am offered would be a wrong one. If it is a good and godly picture, then it is so for someone at a different level, who must use it in his life in a way completely different from anything that would be possible for me" (*CV*, 32e, emphasis original).

> one feels like it but a *capacity* for experiencing another in a way that can't be renounced.[39]

Autonomous, atomistic thinking lacks rootedness, and this brings to question the possibilities and nature of crystalline, secular reasoning. Language games critique the secular reasoning that seeks freedom from the moral touchstones of tradition and authority, and when we do step into the questions of epistemology, we must remain for Wittgenstein rooted and dependent, whether that dependency be upon the "favour of Nature that one knows something" (*OC*, 505) or experience (*OC*, 169, 170) or trust: "I really want to say that a language-game is only possible if one trusts something (I did not say "can trust something")" (*OC*, 509). The challenge in this scientific age lies in understanding the limited and thin possibilities of knowledge in scientific thinking alone, whereas the favorable possibilities for scientific thinking in all its empirical search for precision may be found within a context such as Wittgenstein's religious way of seeing things (*RW*, 94) that requires belief or a received "world-picture," hence enabling the possibility for any balanced empirical observation or hypothesis (*OC*, 167).

> Instead of "I know it" one may say in some cases "That's how it is—rely upon it." In some cases, however "I learned it years and years ago"; and sometimes: "I am sure it is so." What I know, I believe (*OC*, 176, 177).

What is evidenced here is a prioritization, an *ordo* of authority that precedes proof. Hence Incandela states:

> Justification is inherently retrospective here, as a particular "world picture" formed through reliance on authority "is the substratum of all [Wittgenstein's] asserting." This essentially Augustinian reliance upon authority implies that philosophy will always ask more questions than it answers. Its task remains incomplete if for no other reason than its task remains ongoing, just as anyone with children knows that one never *finishes* cleaning the house.[40]

As a way of drawing these strands together, we can conclude that Wittgenstein's religious thinking is a firm step beyond scientific and secular reasoning. Wittgenstein's thought does not move in a systematic or genealogical way, and so his puzzles, aphorisms, and "practices of language" methodology can pose a hermeneutical challenge. For our purposes

39. Asad, "Thinking about Tradition," 167–68 (emphasis original).

40. Incandela, "Augustinian Reading of Aquinas and Wittgenstein," 42 (emphasis original).

here, however, calling upon Wittgenstein's "assembly of reminders for a particular purpose" (*PI*, 127) presents a critique of scientific theorizing that helps to uncover the striking lack of foundation for ethics or bioethics based in theory. Bioethics remains deeply confused when seeking to deduce concrete moral claims when the nature of moral concepts themselves is unclear; the contradictions of modern moral philosophy arise time and again as various theories swirl amidst a fundamental confusion over the possibility of right and wrong in the first place. Moreover, the steady insistence of Wittgenstein's generalist, religious point of view (*RW*, 94) discussed here aptly draws his critique of theoretical philosophy into the terms of discourse between secular and religious ethics. For every challenge Wittgenstein presents to the crystalline purity of logic, the formal unity of sentences and language (*PI*, 107), the scientific precision of explanation (*PI*, 108) and more, we can discern through these gestures his sincere conviction that some other mode of epistemology—and certainly of ethics—beyond discursive rationality is needed to secure something meaningful. Wittgenstein's writings are peppered with hints of his certainty in religious, psychological, even noetic modes of moral knowing. On this point, Monk draws attention to the character of Fr. Zossima, who seems to fit neatly Wittgenstein's notion of psychological insight:

> It was said by many people about the elder Zossima that, by permitting everyone for so many years to come to bare their hearts and beg his advice and healing words, he had absorbed so many secrets, sorrows, and avowals into his soul that in the end he had acquired so fine a perception that he could tell at the first glance from the face of a stranger what he had come for, and what he wanted and what kind of torment racked his conscience.[41]

In a conversation with Drury on Fr. Zossima, Wittgenstein remarked:

> Yes, there really have been people like that, who could see directly into the souls of other people and advise them. Now what would really have interested me would be to have seen how a character like Smerdyakov could have been saved rather than Alyosha.[42]

The latter half of Wittgenstein's remark on Smerdyakov may or may not be taken with any reference to Wittgenstein's admitted inability to loosen his knees so as to pray (*CV*, 56e), but what is clear in the reference to Fr. Zossima is Wittgenstein's insistence on "imponderable evidence" as

41. Monk, *Ludwig Wittgenstein*, 549.

42. Rhees, *Ludwig Wittgenstein*, 123.

an example of the need to go beyond attempts at universalizing rational norms. Wittgenstein's imponderable evidence challenges secular, cosmopolitan rationality by appealing to a variety of human faculties that Wittgenstein lists at random, seemingly at a loss to put words (*PP*, 356) on such an "other" nontheoretical mode or modes of knowing. "Imponderable evidence includes subtleties of glance, of gesture, of tone" (*PP*, 360), and it can be gleaned not through a "course of study" but through *experience*: "What one acquires here is not a technique; one learns correct judgments. There are also rules, but they do not form a system, and only experienced people can apply them rightly. Unlike calculating rules" (*PP*, 355). As Monk implies, abstractions and generalities, laws and principles, and the like that stem from theorizing and intellectualism are staunchly opposed to imponderable evidence that defies the categorizations involved in theorizing norms.[43] "Hence also there can be no ethical propositions. Propositions cannot express anything higher" (*T*, 6.42).

In short, what Wittgenstein offers is a critique of scientific thinking, and this critique of scientism carries with it an essential critique of secular thinking. From within the bounds of a scientific culture, the relationship between scientism and secularism may not be apparent, but in one sense, an implication of Wittgenstein's attempts to practice "religious thinking" may be seen as a Wittgensteinian attempt to look beyond *both* the scientific and secular. When secular thinking is understood as crystalline, cosmopolitan, discursive rationality as devoid as possible of ties to particularity, embodiment, or to concrete communal practices, we have entered a space that admits tradition-full cultures and communities—as well as sincere religions and theologies—only as "private" values not to be admitted into the scientific norms that govern secular space. Wittgenstein finds this impulse confusing and pathological, and bioethics would do well to trace his gestures that point to a more fitting method and way of morality. The norm in bioethics is to bracket off metainquiry in favor of accepting and existing within the theoretical boundaries of modern moral philosophy when the foundations for this discipline itself remain awash in the uncertainty of postmodernity. Wittgenstein's religious thinking pushes us beyond the secular and beyond the reduction of morality as a series of moral calculations on the level of common morality and theories. This is a grave challenge to the *form* of ethics and bioethics.

43. Monk, *Ludwig Wittgenstein*, 548.

Ethics as Apophatic Knowledge and Asceticism

> *It is clear that ethics cannot be expressed. Ethics is transcendental. (Ethics and aesthetics are one.)* (*T*, 6.421)

> *The whole sense of [the Tractatus] might be clearly summed up in the following words: what can be said at all can be said clearly, and what we cannot talk about we must pass over in silence.* (*T*, 3)

Building upon Wittgenstein's aesthetic, ethical, and religious way of seeing the world leads us to the explicit interplay of Wittgensteinian ethics as apophatic knowledge and asceticism. To be clear, it would be misleading to assert that apophasis arises from Wittgenstein as an explicit theological category or dogma. Wittgenstein does not write as an academic theologian examining concepts traditionally rooted in Byzantine philosophy and theology, but rather the aim here is to read Wittgenstein honestly in a philosophical and theological way. Following Drury, Kerr notes that Wittgenstein's religious point of view includes a deep caution for "saying no more than we know," and in this vein Kerr describes the *Tractatus* as "an ascetical exercise in learning to acknowledge what may be said in order to respect that which is unsayable." Hence Wittgenstein's work is "a discipline of reticence," and this duty of apophatic silence possesses a strong ethical dimension.[44]

Additionally, as Clark Carlton has noted, Wittgenstein's conception of philosophy is both therapeutic and apophatic—or more aptly, "it is therapeutic because it is apophatic."

> This is the essence of apophaticism. It is not denial for the sake of denial, or even intellectual humility for the sake of humility. Its purpose is to redirect our attention back toward the

44. Kerr, *Theology after Wittgenstein*, 37. To discern apophatic silence in Wittgenstein is not "the sum" of Wittgenstein, per se, although we may note that such silence occupies a key place following the exercises of the *Tractatus* and *Investigations*. Similarly, Kerr is hesitant to follow Drury to say that the entirety of Wittgenstein's philosophy is "to show once and for all and conclusively that what at first begins as the desire for a metaphysical theory is really something deeper, something which can only be satisfied by other than speculative constructions" (38). Kerr reminds us of Wittgenstein's "endless remarks on philosophical psychology" as an example of philosophical constructions in Wittgenstein. This seems fair, and we need not read Wittgenstein in an unduly monolithic way. As discussed earlier in this chapter with reference to language games as exercises akin to Aquinas's *theological* (vs philosophical) Five Ways (Incandela, "Augustinian Reading of Aquinas and Wittgenstein," 32), Wittgenstein's primary concern seems to be the context and need for authority with such religious thinking. In this case, examining an apophatic methodology as an integral way to read Wittgenstein (see Mitralexis, "Apophatic Wittgenstein") seems complementary to Kerr's primary point on Wittgenstein's post-Cartesian theological turn.

demands of the *Lebensform* by reminding us to play the rules of the grammar.[45]

Carlton's broader point is to assert that the *Lebensform* is important to understanding, for example, St. Mark the Monk's engagement with the fifth-century Nestorians who persisted in asking *how* God does things. Nestorius's questions were "impertinent" and suggest a lack of belief in God arising from a life-form inconsistent with obedient faith. As Carlton employs it, *Lebensform* is not reduced to a physical life pattern in opposition to discursive reasoning—arguably an extension of mind-body dualism—but rather *Lebensform* surpasses such a dichotomy by presenting St. Mark's life-form as the impetus and embodied point of reference for the questions he asks in the first place.[46]

Mitralexis is another who sees Wittgenstein's "stance" toward knowledge and language as apophatic, and in the following I am indebted to a number of the authors Mitralexis has drawn together to explore the relationship between Wittgenstein's analytical approach toward philosophy and "the inherently apophatic nature" of his "*epistemological stance*."[47] As Mitralexis notes, this approach may be called an "apophatic Wittgenstein"

45. Carlton, "Wittgenstein, Apophasis, and Therapeutic Critique," 3.

46. Carlton identifies two sources of heresy: the "false prophet" or "enthusiast" who claims a special relationship with God (e.g., Montanism) and the misplaced "questioner" such as Nestorius. Addressing this latter example, Carlton argues "that most of the major 'isms' that have afflicted Christianity over the last two thousand years, from Sabellianism to Arianism to Nestorianism to Calvinism, began in precisely this way. The need of some *homo sapiens* to construct an ideological system to justify their lives is probably as old as human culture itself. The Greeks were afflicted with it, certainly, and there has been no shortage of such afflicted folk throughout the Christian dispensation, but it was not until the modern era that this attitude became the dominant ethos of the entire 'civilized world.' Indeed, I would argue that this is the hallmark of modernity and that sham 'post-modernism' with its alleged 'incredulity toward metanarratives' has done nothing to significantly alter this ethos" (Carlton, "Wittgenstein, Apophasis, and Therapeutic Critique," 4).

47. See Mitralexis, "Apophatic Wittgenstein," 9 (emphasis original). Mitralexis proceeds to draw from Christos Yannaras's definition of apophaticism in *Relational Ontology* as an *epistemology*: apophaticism is "(1) the denial that we exhaust knowledge in its formulation; (2) the refusal to identify the understanding of the signifiers with the knowledge of what is signified; and (3) the symbolic character of every epistemic expression: its role in bringing together atomic experiences and embracing them within a common semantic boundary marker, a process which allows epistemic experience to be shared, and once shared to be verified" (Mitralexis, "Apophatic Wittgenstein," 9). Further, this definition of epistemology that flows from relational ontology does not contradict the earlier discussion of Wittgenstein's language games as a moral practice more than epistemology; epistemology in the bare sense of "how one knows" is different from the embodied epistemology Mitralexis discusses; such an embodied epistemology *is* moral.

or a "Wittgensteinian apophaticism," and it seems fair to allow either phrase as long as we understand in this approach two key points. First, the *via negativa* employed here is not a deflecting strategy for epistemological questions in ethics. Hopefully a persuasive case has been made by now that Wittgenstein's mysticism is not a failure to address the important aesthetic and ethical questions of life but rather an observation that Wittgenstein insists on noting the limits of scientific theorizing while attending to these questions. A move beyond scientific thinking is not a move into irrationality but may be a move into suprarational religious thinking, and such a higher, mysterious mode of thinking is needed to approach infinite reality. Hence Wittgenstein's postscientific thinking embraces mythology and religion as more fitting ways to encounter transcendent meaning, and we can reflect upon Plato's divided line (*Republic* IV, 509D–513E) an example of the noetic faculty required to comprehend the intelligible things beyond the senses. If we place a hard emphasis on the line between Plato's visible and intelligible, we mistakenly tend toward a mind-body dualism, and yet the noetic faculty that rightly allows one to perceive the intelligible moves beyond such dualism. Hence intellection (*noesis*), understanding (*dianoia*), belief (*pistis*), and conjecture (*eikasia*) present modes of knowing that call for mystery not as an *inability* to answer Wittgenstein's aesthetic or ethical questions but rather as a *higher* ability of suprarationality. Apophaticism that directs one toward meaningless silence is not helpful. Hence the goal for an ethics willing to heed Wittgenstein's scientific and theoretical critique is not inactive stupor, but rather the goal is to identify a suitable mode of being and thinking and then begin the hard work of practicing and living out such being and thinking.

Second, the goal with Wittgensteinian apophaticism here is tied to practices of life and ethics. The argument was made earlier that Wittgenstein's exercises and metaphilosophy follow along the lines of phenomenology, and building upon this phenomenology Sivrides argues:

> Once we make life itself, i.e., action, the object of our thought we are confronted with its limits, meaning that we cannot explain exhaustively the object of our thought: we cannot say what it is definitely. Or, to be more exact, there is something we can say about it and some other thing to which we cannot latch on, something we cannot fetch from our heart to our tongue. Yet we couldn't see the latter if we wouldn't try to seize it. Moreover, if there is a *cause* to strive to understand the state of things, it must lie on the side of the ineffable. Here, we encounter what is known as the theological work of Dionysius the Areopagite, Maximus the Confessor and Maimonides as "apophatic" methodology.

> God cannot be known save through the effect of His activities (*energia*) on our world. Therefore, in apophatic theology the first name of God is that of *summum bonum*, of the sovereign good, denoting ethics instead of cosmology.[48]

Apophatic knowledge articulated in this way is entirely appropriate for understanding Wittgenstein. To misuse language for Wittgenstein is to transgress the limits of scientific thinking by seeking to claim a type of ineffable knowledge that transcends qualitative and causal theorizing. For example, in Peterson's psychological perspective, this boundary between modes of thinking can be seen every time the human mind and soul contemplates anything that extends beyond the mind's calculating capacity. Imagining our solar system, the bright stars in the dark night, or even the layers of another person's emotions and psyche, we are quickly overwhelmed and unable to articulate the entirety of these phenomena. In the face of such scientific limits, Peterson sounds remarkably similar to Wittgenstein by calling for an entirely different mode of thinking. Within the scientific way, the world is viewed as a place of things—the "objective world"—but humans are self-conscious and live amidst narratives—the "world of value." This second way allows us to see the world as a "forum for action and emotion" that is rightly represented through myth, literature, drama, and poetry. Moreover, within narrative, the "unknown" is essential to the unfolding of any story and to reject the unknown as unknowable and hence nonexistent is to exhibit Luciferian pride, which states: "all that I know is all that is necessary to know."

> This pride is totalitarian assumption of omniscience, is adoption of "God's place" by "reason," is something that inevitably generates a state of personal and social being indistinguishable from hell. This hell develops because creative exploration—impossible without (humble) acknowledgment of the unknown—constitutes the process that constructs and maintains the protective adaptive structures that give life much of its acceptable meaning.[49]

One of Peterson's primary aims is to understand myth *not* as "primitive proto-science" but rather as a qualitatively "different phenomenon." The mythic universe is not a place to conceive but rather is a "*place to act*" amidst categorizations of profound value, of what should be. "Action presupposes valuation, or its implicit or 'unconscious' equivalent. To act

48. Sivrides, "Wittgenstein and Kierkegaard's Ethics," 108–09 (emphasis original).

49. Peterson, *Maps of Meaning*, 14.

is literally to manifest preference about one set of possibilities, contrasted to an infinite set of alternatives."[50] Here it is within the tension between finitude (human rationalizing) and the infinite (transcendent reality) that action becomes the embodied focal point of reality, and this is how Wittgenstein sees ethics and religion in the same way:

> This running against the walls of our cage is perfectly, absolutely hopeless. Ethics so far as it springs from the desire to say something about the ultimate meaning of life, the absolute good, the absolute valuable, can be no science. What it says does not add to our knowledge in any sense. But it is a document of a tendency in the human mind which I personally cannot help respecting deeply and I would not for my life ridicule it. (*LE*, 44)

Extrapolating from this, Sivrides makes the point that ethics and religion for Wittgenstein are actions of faith not as irrationality but supra-rationality. Wittgenstein is not interested in the "refined, aseptic God of the philosophers" but rather the "God of the everyman" who is personal and relational. "The way you use the word 'God' shows not *whom* you mean—but instead *what* you mean" (*CV*, 50, emphasis original). Moreover, as with the friend, spouse, or person who is "other" and perpetually knowable yet never defined, God is known personally. He is known definitely but never *definitively*, and this is simultaneously a positive affirmation of God and an apophatic reverence:

> For instance, to designate some action as its very image, namely its project describing its means and end (i.e., how to adjust the appropriate means to an end), is an affirmative (kataphatic) way to define action. Yet saying that the very action is not its image means that it is something more than that. Then the image of an action is shown before us as an artifice that occasions action by making it possible. The scope of the first, kataphatic method is scientific and impartial concerning the knowledge the agent must be equipped with and the calculations he must perform. On the other hand, the scope of the second, apophatic one, is ethical and personal concerning how one chooses and endures the outcome of his choice. It complements the first method by underlying it. Thus the apophatic part is the most important one among the two since that very action can lead the agent to an end that he had not foreseen.[51]

50. Peterson, *Maps of Meaning*, 21 (emphasis original).

51. Sivrides, "Wittgenstein and Kierkegaard's Ethics," 121–23.

Sivrides seems to have in mind an architect's blueprints as an example of the scientific-kataphatic method that calculates the spaces within a house, and these blueprint images are something of a representation of the active lives of those dwelling apophatically within the house. The blueprints indicate—but incompletely, with inevitable limits—the lives within the house. This latter mode is simultaneously religious, apophatic, and ethical, and this latter mode is "more important" for the simple reason that the house is to be inhabited rather than abstracted and conceived.

An important consequence of Wittgenstein's apophatic vision is to pave a path for ethics to be reunited with knowledge and rationality. This knowledge and rationality is premodern in that it is not reduced to scientific categories but remains noetic and religious, and this hierarchical rationality restores ethics as an embodied practice of knowing. Within the scientific paradigm for ethics, we are abandoned and awash within the wake of modern moral philosophy that seeks "moral norms" while ignoring the problem of morality itself that Nietzsche, Anscombe, and others rightly identified. The modernist assumption that moral language may accord in some isomorphic way with moral "reality" remains a spurious claim, and Wittgenstein's practices with language recall us to a standpoint of embodied, social, and historical touchstones. Language and reality *do* connect within bodies, societies, and history as the unfolding of creatureliness, and contrary to the reductive positivist readings of Wittgenstein as one who constructs metaphysical meaning *from out of* immanent practices, we rather should read Wittgenstein as one who sees immanent practices as a way to uncover ontological, transcendent reality *within* and *through* immanent reality. In the former, the active life is an immanent source; in the latter, the active life presences the ineffable transcendent within the immanent.

When the ineffable is discerned within the immanent, the discursively rational qualities of the mind may be humbled and balanced as our attention turns to the epistemology of embodiment, and in so doing we are elevating action, habit, and practice beyond nonintellectual materialism.[52] Further, considering language as a parallel to embodiment, Kerr notes that in speaking of God we must take care how the word "God" is used. This apophaticism recognizes layers and modes of seeing God that engage the thinking, acting, worshipping man:

52. Here one might benefit in drawing connections between Ravaisson's phenomenological embodiment in *Of Habit* and Wittgenstein's phenomenological language practices. Further, as Foltz has argued, a parallel may be drawn between Husserlian phenomenology and the "noetics of nature." From this perspective, material nature may be seen radiating "the beauty and goodness of creation (*ktisis*) whose infusion with divine glories is now becoming manifest (Folz, *Noetics of Nature*, xvii).

> Faith, in appropriate circumstances, is visible in one's behaviour; it is not some undetectable inner object. . . . Language, the living human being, our life, human nature: Wittgenstein's watchwords in the philosophy of psychology are also contributions of central importance to a theology that starts where we are; a theology for ceremonious animals, so to speak, rather than for cerebriating solipsists; a theology that starts from the deep and sinister thing in human nature rather than a hypothesis about a deity; theology *naturalized*, so to speak.[53]

Building upon this view of theology *naturalized*—so clearly juxtaposed to natural theology—Kerr turns to *PI*, 610, where Wittgenstein ponders describing the aroma of coffee. "Why can't it be done? Do we lack words? And *for what* are words lacking?—But whence comes the idea that such a description must after all be possible?" (emphasis original). The point is not that there are no words to describe such an aroma but rather that words are limited and insufficient. It may seem a small point and mere "common sense" to accept our limited ability as the ability we need, and that is that, but the process of noting, accepting, and exploring these limits is certainly worth our attention. The aroma of coffee *can* indeed be described and yet the kinds of description we employ inevitably fall short of experiencing the aroma personally, and that personal experience remains the important thing.

Furthermore, Kerr suggests another comparison as Wittgenstein imagines trying to count the number of falling raindrops in a storm. As one assesses the limits of language, is it correct to become dissatisfied? If so, is one also dissatisfied with the limits of vision in not being able to count the raindrops with "exactitude or completeness?" Such a comparison seems "deliberately crude," and the effect is to startle us into "realizing that we do have an ideal of exactitude or completeness at the back of our minds which very easily imposes itself inappropriately."[54] These illustrations and comparisons, moreover, therapeutically draw us into an awareness of our finite engagement with the infinite reality that surrounds us on all sides, and this infinitude always surpasses crystalline, scientific knowing. Music or poetry may be rightly analyzed via the critical gaze, and, to be sure, an art critic's words have their place as long as the commentary remains clearly distinct from the poetic acts of hearing and imagining. The enjoyment of these arts is akin to reading a facial expression: no one truly doubts the deep meaning

53. Kerr, *Theology after Wittgenstein*, 163 (emphasis original).

54. Kerr, *Theology after Wittgenstein*, 165.

embodied within facial movement acts,[55] but the point is not to possess an *expectation of certitude* that will then lead to despair when words fall short in explaining the face before us. For Kerr, the correct response is "to awaken to the possibility that our way of life is the incomparable thing that it is, without compulsively contrasting it all the time with alien alternatives. The metaphysical picture of what we are, far from securing our uniqueness, only obscures it."[56]

This embodied anthropology and psychology is post-Cartesian in an important way. If we accept Kerr's contention that the metaphysically structured self is a—if not *the*—central concern of Wittgenstein's later writings, then Wittgenstein has much to offer in changing the form of our ethical vision, giving us a rather different expectation and methodology for ethics and bioethics. Perhaps the most important part of this different vision is to accept human finitude not as a flaw to be overcome but as a critical part of the ascetic way of being within the cosmos. Here asceticism is a comprehensive vision of ethics as it places the human not as a "metaphysically structured self" who must operate within a theoretical sphere to calculate correct actions grounded superficially within a shifting political common morality consensus. Instead, the ascetic man is one who lives within the processes of redemption, seeking reunion with the ineffable One. This asceticism embraces human formation without the presumption of perfection that the metaphysical tradition arguably assumes and/or encourages.[57] Hence the Genesis creation story perpetually draws our attention to Adam and Eve refusing to accept "moral humanity" and "creatureliness" as they crave to be

55. Ours is an age that places great emphasis on interpreting "body language" as evidenced in the attention given to poker games and particularly at ceremonious events such as a presidential inauguration, a first dance at the inaugural ball, etc. The "science" of body language became known as kinesics during the mid-twentieth century.

56. Kerr, *Theology after Wittgenstein*, 166. Kerr sees in Wittgenstein a deep engagement with the "story of the soul in the Western metaphysical tradition," and he quotes St. Augustine on "joy" in a sermon on Psalm 32: "At the harvest, in the vineyard, wherever men must labour hard, they begin with songs whose words express their joy. But when their joy brims over and words are not enough, they abandon even this coherence and give themselves up to the sheer sound of singing . . . And to whom does this jubilation most belong? Surely to God, who is unutterable. And does not unutterable mean what cannot be uttered? If words will not come and you may not remain silent, what else can you do but let the melody soar?" (166–67).

57. The metaphysical tradition seems bound to assume a perfect nature somewhere "outside" humans, and this nature is postured as the object and aim of human being. This is an "other" that does not give relational knowledge through intercommunion but rather as an ideal that divides the "true" self from one's self—hence, a relegation of one's self to something less than the metaphysical ideal, whatever that may be. This perfection in theory seems equally bound to direct one towards either existential despair or to individual delusional hubris and grandeur.

"immortal, omniscient, omnipotent." God as divine creator alone possesses these qualities, and the unbounded Creator-beyond-being creates man who is not divine. In this vein Kerr extrapolates upon Schillebeeckx's argument that belief in God is a repudiation of metaphysical dualism:

> The doctrine of creation ex nihilo was historically at the centre of the Church's polemic against Gnosticism (e.g., Irenaeus). For Schillebeeckx, the point of Christian belief in the world's being created from nothing is that it still rules out dualistic metaphysics. When we first reflect on the human condition—"Was I anywhere, or anyone?"—it remains tempting to succumb to the thought that the self is out of place in the world, and then bodiless and language-free experience, epistemological solitude, and so on, at once become unavoidable and even desirable possibilities. Thus we return to the ancient picture of human life as essentially exile from our true nature.[58]

Similarly, drawing further on the aroma of coffee, Sidiropoulou notes that:

> None of us is above space and time. Conditioned as we are by our embodied nature, we are embedded in particular and specific contexts of life that give shape to different shades of similarity in our mental life or, indeed, to completely different experiences. Given this, to ask for a unique and final representation of subjective experience (the aroma of coffee) would equal a demand to revoke human finitude. For only an infinite and, by definition, immaterial being can be above the particularities, the different perspectives, viewpoints and situations of life that render the aroma of coffee so unique for each one of us.[59]

Ethics as asceticism embraces both a subjective and objective aspect as individuals are not united through a cosmopolitan rationality but rather through individual experiences amidst a communal way of life.[60] This way of life, moreover, remains for Wittgenstein an active, ritual, and liturgical experience that allows the embodied person to seek union and to participate

58. Kerr, *Theology after Wittgenstein*, 185.

59. Sidiropoulou, "Wittgenstein's Apophatic Descriptions," 69.

60. By "common way of life" here we may think of the aroma of coffee as an action that can be shared concretely. This concrete experience *itself*, however, is not *the* answer to the Cartesian error. The aim is not simplistically to suggest that such a simple act or way of life itself is the remedy to metaphysical dualism by retreating from cognitivism into praxis. Rather, the aim is to follow a Wittgensteinian therapy to see a more embodied epistemology—the interconnectedness of *theoria* that transcends *scientia*.

corporately within the ineffable transcendent One.[61] The temptation toward perfection in metaphysical dualism leads one to seek transcendental moral categories whereas ascetic embodiment leads toward practices of union with the ineffable transcendent God. We need not see in this emphasis on embodied practices a "crypto-behaviorism" as though religious actions and the "dancing out" of life are aimed at "people's lives and imaginations" more than at the deity to which they refer. Such a concern is "*mutatis mutandis*, the same kind of apprehension as the one regarding the ineffability of the coffee's aroma; a fear of losing one's own innermost self. . . [it is] a fear of having to give up God Himself, through reducing him to the pure immanence of social practice." However, as Sidiropoulou notes, this collapses again into a dualistic metaphysics of ineffable presence when one assumes that "God" (a sort of an object or concept) logically precedes any embodied, liturgical response to "it." Liturgy in this way is linked as a "causal outcome" to the concept of God rather than as an "internal link" or deep union. In short, for those who see liturgy as an outcome of God, "practice must not be a constitutive part of what is involved in the concept 'God,'" and hence in seeking to avoid a "reduction of the 'beyond' to external acts," the way of dualistic metaphysics is reopened.[62]

Sidiropoulou suggests that critics of Wittgenstein miss the point with the aroma of coffee when they seek an esoteric *description* for experiences-beyond-words or when they seek to *picture* exactly what is invisible.[63] Wittgenstein's apophaticism instead directs us to see that human description or pictures are *not* in fact able to contain adequately ineffable transcendence. Does this then imply that the *via negativa* lacks any *kataphasis*, and is there an understanding of the ineffable that is nondualistic and nonmetaphysical?

61. This broad vision of asceticism is very much in line with the traditional Christian use of *paideia* as wisdom. "The Old Testament uses and develops the theme of wisdom in a number of ways: as empirical knowledge and intuitive understanding, as formation or *paideia*, as cosmological principle and mediator, as active divine subject, and as a means of expressing God's transcendence and immanence. In looking at the patristic understanding of wisdom we also need to keep an eye on the classical hinterland. While far from being a univocal concept in classical antiquity, wisdom, with its cognate terms, rapidly emerged as one way of describing the character of the enquiry into the nature and meaning of things. The limits to this enquiry were soon recognized but this sense of the fallibility of human conclusions was to some degree balanced by an appeal to the divine source of such wisdom and knowledge as humans might possess. This explicitly theological approach was also not infrequently bound up with a mystical and moral/ascetic dimension. . ." (Plested, "Wisdom in the Fathers," 240).

62. Sidiropoulou, "Wittgenstein's Apophatic Descriptions," 71.

63. Here we may contrast "pictures" that result from capturing some visible thing with "images" (*eikon*) that draw one in to contemplate through the illumined visible image a higher reality beyond bodily vision.

The answers are "no" and "yes," respectively, for as Wittgenstein describes it when directly addressing ethics, both finitude and the ineffable are present, though present as a paradox. As the "Lecture on Ethics" intimates, it is an ethical paradox that is humbly kataphatic, embracing finitude, and yet when Wittgenstein speaks of yearning "to go beyond the world" and "say [what is] beyond significant language" he is describing the deep tension between the kataphatic *and* apophatic, and "the whole tendency of all men who ever tried to write or talk about ethics or religion was to run against the boundaries of language. This running against the walls of our cage is perfectly, absolutely, hopeless" (*LE*, 18–19).

Moreover, these creaturely limits in our morality call for a *persistence* in practice and way of life while never ignoring the eschatological union of future and present (Heb 2:8–9; Eph 2:6). On the one hand, as described above, such a paradox or antinomy should be taken as Wittgenstein's critique of culture and an attempt to challenge "the pervasive disease in the culture of his times, namely, the obsession with scientific progress." This entails that Wittgenstein's apophasis challenges the "spirit" underlying philosophy in the modern era that is tied to the way people live (*Lebensweise*).[64] On the other hand, beyond Wittgenstein's cultural application, Sidiropoulou directs us to St. Gregory of Nyssa's *Life of Moses* (II.3) to understand how one may balance the tension between practices of *going beyond the world* (*LE*, 18) while admitting the limits of language. For St. Gregory, the unity of this paradox may be seen as the soul—the beloved moved toward the Lover—is drawn into a loving participation with the divine source of all being. From the moment the soul glimpses the loveliness it sees to when it is drawn into that transcendent love, we glimpse a participation like that of "an image" (*eikon*) that is fulfilled in "the archetype." As St. Paul writes in I Corinthians 13:12, we move from seeing dimly in a mirror to then seeing face to face. Hence while we are perpetually "limited" in our embodied and social being, we are nevertheless enabled to see through mirrors and "representations of God as the object of our desire."

> These representations we share with others, for after all, they arise within our common life and in interdependence with our culture. A representation of God is not elusive in the Cartesian sense of conceptual privacy. Where it remains elusive, though, is in its radically "incomplete" character. Even the use of the term "incomplete" here is regrettable for it seems to suggest that a completion may be possible.[65]

64. Ip, "Identifying the 'Apophatic Impulse,'" 101, 100–05.

65. Sidiropoulou, "Wittgenstein's Apophatic Descriptions," 73.

Moreover, in terms of language, this process of representation is both affirmative and negative:

> Human language has the means to talk about God and has been doing so for thousands of years. Our words, expressions, and concepts of the divine are fine as they are. This means: they are not incomplete and second best in lack of better ones that could offer an all-encompassing representation of God. For such a thing to be possible, *not only God but also we* would have to transcend our human embodied condition and our creaturely nature. *Such an effort, as Gregory of Nyssa straightforwardly warns us, borders idolatry.*[66]

A further key point regarding the interplay between the apophatic and kataphatic is to avoid contrasting simplistically the divine darkness with the divinely dazzling light, or through a sleight of hand to enact a simple "positive" from out of a "negative." To misjudge the tension between apophatic and kataphatic as discussed here is akin to reading Wittgenstein's language games as a strategy for enacting metaphysical immanence. In both cases, respectively, a practice of uncovering humble limits with discursive language does not automatically "create" light out of darkness. Instead, the practice of negation is *formative* and even liturgical, whereby there is no causal moment of separation between negative and positive, but rather a seamless movement of the finite, loving soul into infinite love. Hence in the way that Dionysius addresses divine names, the truth is sought *through* the divine names of the sacred Scriptures not on the terms of plausible human wisdom but by the power of the Spirit granted to the writers of the Scriptures. This employment of language moves beyond language into a "union superior to anything available to us by way of our own abilities or activities in the realm of discourse or intellect."[67] Dionysius maintains that the deity is invisible and incomprehensible as well as unsearchable and inscrutable while also not being "incommunicable" to everything.[68] What is required is a movement of the entire human being to be raised "firmly and unswervingly upward into the direction of the ray which enlightens them. With a love matching the illuminations granted them, they take flight, reverently, wisely, in all holiness."[69]

66. Sidiropoulou, "Wittgenstein's Apophatic Descriptions," 75–76 (emphasis original).

67. Pseudo-Dionysius, *Divine Names*, 585B.

68. Pseudo-Dionysius, *Divine Names*, 588C.

69. Pseudo-Dionysius, *Divine Names*, 589A.

Hence the dynamic interplay between negation and affirmation is a working of "divine enlightenment" (*theurgy*),[70] and the Dionysian approach for speaking of God is first to employ analogies and symbols.[71] Moreover, following Him Ip's suggestion that apophasis is a hermeneutical key to unlocking Wittgenstein's *Lecture of Ethics*, we can easily see Wittgenstein resonating with Dionysius's language on "climbing higher" to the supreme, ineffable Cause of conceptual beings Who is not himself conceptual:

> There is no speaking of it, nor name nor knowledge of it. Darkness and light, error and truth—it is none of these. It is beyond assertion and denial. We make assertions and denials of what is next to it, but never of it, for it is both beyond every assertion, being the perfect and unique cause of all things, and, by virtue of its preeminently simple and absolute nature, free of every limitation, beyond every limitation; it is also beyond every denial.[72]

In particular, the notion of making "assertions and denials of what is next to it, but never of it" seems a remarkably apt description for Wittgenstein's philosophical practices. He remains revered by many for his analytical insights, and yet for Wittgenstein there was little joy in "how little had been done" as he set out to "throw away the ladder" after he had climbed it (*T*, 6.54). Wittgenstein hence gestures and speaks of *das Mystiche* beyond his assertions and denials of *what is next to it*, and one is left with something of a "what if" had Wittgenstein sustained further his deep interests in the mystical reality that he sensed beyond his clarifying practices. Dionysius in particular would have seemed a fitting starting point for Wittgenstein given his theological call to the *via activa* in liturgical practice. Whereas many read Dionysius as a philosopher, the stronger case is to read him liturgically as he calls on the divine names to end in "hymns" of praise, and this ecclesial, liturgical context serves arguably as the touchstone for all his works.[73] Indeed, in *Mystical Theology* I.3, Moses's ascent mirrors Dionysius's threefold theological methodology of affirmation, negation, and ascent into darkness, and within this darkness is a union beyond mind, knowing and unknowing. Who is the likely audience for this except the "laity" as those who contemplate the liturgical acts within which they are bound? Within this liturgical way, we can glimpse something beyond an intellectual exercise but more of a sculpting of persons where the stripping away of something (knowledge) is not simply negative but reveals instead

70. Pseudo-Dionysius, *Divine Names*, 592B.

71. Pseudo-Dionysius, *Divine Names*, 592C. See also *Mystical Theology*, 1033A–B.

72. Pseudo-Dionysius, *Mystical Theology*, 1048A–B.

73. Pseudo-Dionysius, *Ecclesiastical Hierarchy*, 429.

a higher, inner beauty that only the sculptor initially imagines. In short, for Dionysius, the apophatic and kataphatic are not sequential but rather held together, and the consummation of his theology points to an active, *theurgic* life.[74] God is not chiefly to be "thought of" or "gazed upon" but rather called upon and worshiped.

Some may contest at this juncture that the apophatic theology of Eastern orthodoxy is particular in a way that belies Wittgenstein's legacy. To this, little needs to be said.[75] While Wittgenstein promotes the directionality

74. Indeed, the *theurgic* life, or *theosis*, avoids the trappings both of immanentism and radical transcendence. In the postscript to *Theology after Wittgenstein*, Kerr responds to Russell Reno's concern that "Christian attempts to express the delicate balance (or violent tension) of the ordinary and the extraordinary in the life of pilgrimage always fall by the weight of their concepts and the inner desire for escape into . . . radical transcendence" (Reno, *Ordinary Transformed*, 159). Reno's concern is that Kerr is "so suspicious of the language of transcendence" that Kerr has given rise to a "spiritual extrinsicism in which ordinary life becomes separated from the goal and aspirations of God's extraordinary offer of love" (199). Kerr generously allows for such a critique while pointing to the tension of being human by quoting *Nichomachean Ethics*:

> We should not heed those who counsel us that, being men, we should think human, and being mortals, we should think mortal. But we ought, so far as is in us lies, to make ourselves immortal, straining every nerve to life in accordance with the highest thing in us (1177b 31–34; Kerr, *Theology after Wittgenstein*, 189).

Further, within the mystical path of the Egyptian ascetics, *theurgy* as embedded within *katharsis* (purification of the soul), *theoria* (contemplation of the divine energies), and *theosis* (divinization and union with God) presents a way to move beyond the tensions of immanentism and radical transcendence. It is not possible here to do more than allude to the rich theology of *theosis*, but it is helpful as Lossky notes to see the "objective nature of grace" as one aspect of the church, particularly manifest "in the sacraments, in *theurgic* actions, the hierarchy and the power that it wields, the Church's worship, the sacred symbols. . ." This objective nature illustrates the christological aspect of the church, and interwoven to complete this objective aspect is the pneumatological aspect that is "personal." The christological aspect is a more "external and functional presence of grace" whereas the pneumatological is "appropriated by the person, becoming *personal*." Together, these two aspects undergird the harmony and union within *theosis* (Lossky, *Mystical Theology of the Eastern Church*, 190–91, emphasis original).

75. To be sure, there are other nonscientific or premodern visions of the world that may guide the moral life, and the approach taken here with Wittgenstein opens the door to a variety of religious ways of life. A central aim of this work has been to follow Wittgenstein therapeutically beyond the trappings of scientific thinking to the threshold of next-level questions. Here many may agree that a religious view of the moral life within a divinely ordered cosmos is true and then disagree strongly on Who the divine God is. Jews, Christians, Muslims, Buddhists, Confucians, and more may find Wittgenstein's ascetic moral vision helpful amidst the twenty-first-century theoretical, scientistic *zeitgeist*. In this way, Wittgenstein paves the way for rediscovering the necessity of premodern, religious thinking in a secular, scientific world. His tactic is postmodern in its critique of immanent language, and yet his postmodernity is strategic as his greater allegiance lies with the premodern view of philosophy as a way of life.

of apophatic thinking, he never converted to Orthodox Christianity, even though conversion would be a familiar and fitting premodern category—a category within, say, Hellenistic or Roman schools of philosophy—for describing the nature of changing one's philosophy or ethics as "spiritual exercises." Hadot speaks of a transformed ethics as spiritual exercises that

> correspond to a transformation of our vision of the world, and to a metamorphosis of our personality. The word "spiritual" is quite apt to make us understand that these exercises are the result, not merely of thought, but of the individual's entire psychism. Above all, the word "spiritual" reveals the true dimensions of these exercises. By means of them, the individual raises himself up to the life of the objective Spirit; that is to say, he re-places himself within the perspective of the Whole ('Become eternal by transcending yourself').[76]

So, while Hadot differentiates analogically his generalized assessment of spiritual exercises from asceticism such as that practiced within traditional Christianity, his premodern stance and analysis of ancient philosophy as a "supraethical" practice is helpful for changing the form of ethics. Such a turn, moreover, brings us full circle to the first chapter of this book where Jahr and Potter's spiritual and cosmic articulation of bioethics challenges Helleger's effective reduction of bioethics into medical ethics solely as an applied project. In contrast, one can see in the Stoics a supraethics of "learning to live" not by values dependent upon our "passions" but upon "the perspective of universal nature." Similarly, the Stoics saw "learning to dialogue" as well as "learning to die" and "learning to read" as self-formation, *paideia*, to guide learners not "in conformity with human prejudices and social conventions [e.g., politically-grounded common morality]. . . but in conformity with the nature of man, which is none other than reason."[77] Such a praxis-rationality is fundamentally embodied as it seeks freedom of the self from "worries, passions, and desires" to "attain to wisdom." In striking contrast to the autonomy approach of euthanasia "rights" in the modern era, these spiritual exercises are examples of what bioethics could do if embedded within the presuppositions of ancient philosophy. Contrary to the fundamental puzzle of modern moral philosophy and the theoretical calculations within principlist bioethics, Wittgenstein's approach supports philosophy "in its original aspect: not as a theoretical construct, but as a method for training people to live and to look at the world in a new way. It is an attempt to transform mankind."[78]

76. Hadot, *Philosophy as a Way of Life*, 82.
77. Hadot, *Philosophy as a Way of Life*, 102.
78. Hadot, *Philosophy as a Way of Life*, 107.

Furthermore, beyond the Hellenic and Roman schools of philosophy, Hadot points to the "absorption of *philosophia*" by Christianity—especially in monasticism—that stands to reframe ethics even further. It is within this spirit that Engelhardt unapologetically presents "Christian bioethics" and the way of life within the Eastern Orthodox tradition as relevant and of interest not only for baptized Christians but for those who wish to see a model of premodern bioethics as a *Lebensform*. Engelhardt's *Christian Foundations of Bioethics* doubtless remains provocative to those who embrace a cosmopolitan progressive vision of reality—the cross remains a *stumbling block* and *foolishness* (I Cor. 1:23)—yet one need not be a Christian to discern the stark contrast between Christian bioethics and the immanent theorizing of modern bioethics. In these ways—aligned with either Hadot's spiritual exercises or Engelhardt's communitarian bioethics—Wittgenstein's ethics persist in reminding us of the inevitability and importance of attending to *methodology*. Rather than resorting to ethical visions that in effect reduce all of reality within the limits of scientific thinking, Wittgenstein's ethical praxis is a therapeutic exercise aimed at nothing less than the transformation of the person.[79] Wittgenstein does not seek to create a list of proper applied ethical actions but rather offers a corrective meditation on the ethical person who lives—in *bios* and *vios*—within both sides of the tension between immanence and transcendence, finitude and infinitude. In this regard, it may be that Wittgenstein's most pertinent contribution to bioethics is the inexorable reminder that there can be no bioethics in abstraction without concrete, moral human persons who are acting and experiencing beings amidst a world of wonder.

In this way, we must recall that such an ascetic and apophatic ethical vision necessarily remains distinctly personal. It is not focused on external coalitions or factors that can be easily framed in universal formulas; hence if the modern world's reduction of ethics to theory is accepted, Wittgenstein has no ethics. The same may be said of Orthodox Christianity. A premodern ethical vision, however, is particularly internal, and paradoxically this

79. It is notable in this context that Hadot draws from Düring the insight that Aristotle did not intend to set forth "a complete system of reality" but rather "wished to train his students in the technique of using correct methods in logic, the natural sciences, and ethics." Hadot quotes from Düring on Aristotle to make what can be taken as a very Wittgensteinian observation: "the most characteristic feature in Aristotle is his incessant discussion of problems. Almost every important assertion is an answer to a question put in a certain way, and is valid only as an answer to this particular question. That which is really interesting in Aristotle is his *framing of the problems*, not his answers. It is part of his method of inquiry to approach a problem or a group of problems again and again from different angles" (Düring as quoted in Hadot, *Philosophy as a Way of Life*, 105; emphasis added).

internality should also be seen in a cosmic way.[80] Solzhenitsyn grasps this by noting that the good and evil of the world are a matter of personal responsibility. That is, Solzhenitsyn discovered powerfully amidst the horrors of the Stalinist Gulag that he possessed personal responsibility for the barbarism surrounding himself. This paradox or antinomy stems from a conviction that "the Universe has as many different centers as there are living beings in it . . . each of us is the center of the Universe . . ." The entirety of the world around is cosmically interwoven within us and we within the world. Morally, this is a vision that is practical not in formulas or procedures but in manifestation of every individual cosmic-moral person. In a way Wittgenstein surely would appreciate, Solzhenitsyn speaks of the mystery of morality that is ineffable yet immanently active and personal:

> If only there were evil people somewhere insidiously committing evil deeds, and it were necessary only to separate them from the rest of us and destroy them. But the line dividing good and evil cuts through the heart of every human being.[81]

From Scientific Ethics to Cosmological Ethos

> *In the language of our culture today . . . Each man's morality or ethos is his objective valuation based on a scale of values accepted by society as a whole. This scale of objective virtues or values which determine the morality of the individual is the core, or the result, of a more general religious, philosophical or rationalistically "scientific" interpretation of problems to do with the behavior of the individual in society. Ethics is the name customarily given to this kind of systematic concern with the problems relating to the ethos or morality of the human being. —Yannaras*[82]

The final turn to *ethos* and *cosmos* will, in a real sense, be one that readers will be in step with or not. Those who are comfortable with ethical theories and social conventions of immanence as the grounding for morality will find little value in exploring how the pre- or postmodern world challenges such a scientific ethics. Similarly, those who embrace the philosophical exercises in positivism the early Wittgenstein mastered—and then abandoned—as meaningful epistemology will not be moved by the later Wittgenstein who embraces qualities of life such as apophaticism,

80. Peterson, *Maps of Meaning*, 336.

81. Solzhenitsyn, *Gulag Archepelago*, 13, 168.

82. Yannaras, *Relational Ontology*, 13.

asceticism, cosmology, and mysticism. The differences between these rival approaches here have followed scientific thinking as a sort of phenomena, a phenomena revealing hidden theological maneuvers that are inevitably embedded within science, technology, theorizing, and modern life that so desperately seeks to be objective, neutral, and fact based. Behind every facet of such a modern, cosmopolitan, liberal *Lebensform* can be found a self-supplied pseudotheological vision of self-made goods. In the absence of some imparted, participatory life within the good, we inevitably supply our own. Hence as the Yannaras quote above indicates, modernity seeks an irreligious move to scientific ethics.

These comments are not intended as a sour grapes disposition toward those not in step with the argument of this book. The point, rather, is that it helps no one to ignore the vast divide between the worldview described here and the worldview of modern scientific thinking. In a real sense, this book is not a book in "bioethics" because mainstream bioethics is defined by scientific thinking such that to go "beyond bioethics" may mean to abandon the term altogether, or at least to hold perpetually a critique in mind when maneuvering through the minefield of bioethics dominated by principlism and scientific thinking. Similarly, it does no good to ignore the philosophical chasm between the logical positivists inspired by the youthful Wittgenstein and those inspired by Wittgenstein's turn to mysticism and religious thinking.

In *finis*, a final meditation via cosmology is helpful to summarize this essay in moral theology. Why cosmology? Simply put, beyond encapsulating the core themes of this book, cosmology outlines a way to save ethics from the enslavement of scientific thinking. By cosmology, I mean ancient and sacred cosmology as described well by Philip Sherrard's understated, yet powerful, *Human Image World Image: The Death and Resurrection of Sacred Cosmology.*[83] First and foremost, cosmology means that our understanding of the universe begins with our understanding of ourselves, and the destiny of the world is essentially linked to our human destiny. Anthropology hence precedes cosmology for "our model of the universe—our world picture or world image—is based on the model we have of ourselves, upon our own self-image. When we look upon the world, what we see is a reflection of our own mind, of our own mode of consciousness."[84]

What this understanding of cosmology offers is a way to see both ourselves and the world we inhabit as reciprocal aspects of a harmonious and shared divine creation. Can there be a more apt description for

83. Sherrard, *Human Image, World Image.*

84. Sherrard, *Human Image, World Image*, 2.

the intersection of *bios* and *ethos*? Parting ways with Descartes, we are not radically divided between *res cogitans* and *res extensa,* where thought and sensible things are discordant; instead human beings are

> descended from the gods, or from God, and as heirs to eternity, with a destiny that goes far beyond politics, society and economics, or anything that can be fulfilled in terms of the mortal world or by satisfying their moral and physical desires and needs. They think of themselves as sacred beings, even as semi-divine beings, not in their own right, but because they are created in the divine image, in the image of God, of a transcendent more-than-human form of consciousness. They come from a divine source, and the divine world is their birth home, their true home.[85]

What this means is that human beings within a sacred cosmology do not see themselves as materialist evolutionary beings, evolved in a meaningless, naturalistic or purely mechanistic formula. Human beings belong *with,* not against, nature, recalling that nature is not something external, fatalistic, nor impersonal but rather is being created within the will of the Divine. In this way, we speak of divine humans *and* divine nature, and neither can be apart from the other. The implications of this cosmology are paramount for science and ethics, for in the same way that transscience oversteps its boundaries, revealing the method of autonomous humans ruling over the world in a technical mode of endless engineering, so, too, ethics can seek to manufacture human behavior within a system of endless rules suspended from a proposed theory. Rather than living freely and harmoniously within a divinely ordered cosmos, modern ethical individuals seek a formulaic or scientific posture. Hence ethics is stifled as the mere "'scientific' interpretation of problems to do with the behavior of the individual in society."[86]

Ethics in such a framework is entirely reductivist, akin to the analysis of created being—the Spirit, *moving* over *the face of the waters* (Gen. 1:2)—solely through physics or numeric abstraction. This view of mathemathics *qua mathematics* or science *qua science* neglects the spiritual and eternal qualities of nature by acknowledging only physical parts of the divine whole, and the ensuing problem is not that science overtly wars against spirituality but rather that we expect real scientific knowledge "as science" to be bracketed apart from the spiritual and eternal aspects of the natural world. For Sherrard, to study the singular, immanent mode of life "extended in time and space" without reference to the higher mode

85. Sherrard, *Human Image, World Image,* 4.

86. Yannaras, *Relational Ontology,* 13.

of reality, the "sphere of spiritual knowledge or of a spiritual science," is a fallacy with dire consequences. Namely, to assume that science is complete "as science" naturally leads to a view of ethics "as ethics" also being complete. What is lost is the deeper knowledge of things and beings as they truly are, existing within multiple modes of reality. Hence the necessity of spiritual consciousness that "perceives and experiences things as they are in themselves, inner and outer, spiritual and material, metaphysical and physical interpenetrating and forming a single unsundered and unsunderable reality."[87] When this unified physical and metaphysical tapestry is sundered, when the scientific mentality birthed from unquestioned assumptions within "medieval scholastic thought" has triumphed, the result is nothing less than the rape of man and nature.[88]

In Sherrard's language, the higher mode of knowing in "spiritual consciousness" that flourished in many ancient civilizations has been eclipsed by "ego consciousness," and we can do nothing but take ego in the fullest sense of autonomous modern rationality that is consistently held up in bioethics and stridently critiqued in the later Wittgenstein. The human faculty for this ego-knowing is human reason (*dianoia*) in isolation and existing in a sort of epistemological *media res*, an in-between state that depends upon a starting proposition or theory. Without such a suspended starting point, there is nothing upon which *dianoia* can operate, for discursive reason is not "self-generated, as though by a process of internal self-combustion."[89] Discursive reason in isolation is no phoenix but rather a suspension bridge that requires a foundation set upon one shore with the hope of somehow extending to the other shore.[90] Without a first premise or Absolute (*arche anypothetos*), reason flails amidst the void. Every act of reason or thought must remain grounded in faith: "[s]uch an affirmation of faith stands at the head of even the most positivistic scientific explanation or demonstration, however elaborate it may be or however elementary and self-evident it may appear."[91] In other words, the limits of isolated discursive reason are nothing

87. Sherrard, *Human Image, World Image*, 8.

88. See Sherrard, *Rape of Man and Nature*.

89. Sherrard, *Human Image, World Image*, 82.

90. Fundamentally, modern science assumes a reliability as opposed to chaos when constructing theoretical knowledge. "The truly apocalyptic view of the world is that things do *not* repeat themselves. It isn't absurd, e.g., to believe that the age of science and technology is the beginning of the end for humanity; that the idea of great progress is a delusion, along with the idea that the truth will ultimately be known; that there is nothing good or desirable about scientific knowledge and that mankind, in seeking it, is falling into a trap. It is by no means obvious that this is not how things are" (*CV*, 56e, emphasis original).

91. Sherrard, *Human Image, World Image*, 83.

short of overwhelming when this mode of knowing alone is presented *as* yielding knowledge. Discursive reason alone most lacks direct experience for it is not grounded in intelligible or spiritual realities; it is not harmoniously *within* but rather something that forcefully works *upon*. Beyond the ego, the higher organ of consciousness that experiences spiritual reality is the *nous* or *intellectus*, and without such a reality we remain bound within the ethics of modernity that Yannaras aligns with rationalism and "scientific interpretations." In this way, ethics via discursive reason focused upon moral problems requires an "agreement or convention." "We accept, on utilitarian grounds, ideals about good and evil which may either be those put forward in each particular case, or those hallowed by experience. And we constantly seek to improve them, using philosophy or science to study the manifestations of social behavior."[92]

The ethical bonds within isolated discursive reason are as thick or thin as current society, convention, or a theory, and these are the shifting sands upon which modern ethics claims knowledge (*scientia*). Science, Wittgenstein notes, is simultaneously "enrichment and impoverishment. *One* particular method elbows all the others aside. They all seem paltry by comparison, preliminary stages at best" (*CV*, 60e, emphasis original). Yet against this monopolizing enslavement of ethics to scientific theorizing, we can contrast a focus on the individual versus a focus on the cosmos as a way to clarify things. The modern individual is one who buries himself within the "conventions of utilitarian law" or the authoritative conventions of "ideological or party authorities," and this offers an escape from his own existential problems. The ontological questions of what man *is* do not lead to what man *ought* to do; instead, a thinner starting point yields a system of palliating the depth of ethics within surface orientations. The inevitability of some hierarchy and authority remains undeniable, yet that authority stems either from a robust ontological commitment of spiritual man within a divine cosmos or to the ego identity of man typically born within utilitarian grounds. The latter source of authority has been the primary focus of this critique, but the former source brings us to a cosmology that roots persons within a harmony of being far beyond the discord of theorizing individuals. As Yannaras summarizes, the question of ethics rooted in ontology is fulfilled by the "ethos of trinitarian communion"; the personal existence of God is the fullest expression of the truth of being.

92. Yannaras, *Relational Ontology*, 14. As discussed in chapter 5, Engelhardt's theory of permission provides a case study of how one may navigate the limits of convention and agreement; the image of a handshake serves to exemplify the thin bond and limited reality when convention and agreement alone hold together dialogue (*dia* + *logos*) amidst thin, mutual commitments to peace as morality.

Not the "essence or the energy" of God but His "personal *mode* of existence: God as person is the hypostasis of being."[93]

Here we must recall that the God of ancient Israel did not manifest himself primarily as a dogma or written message. His chief gift was his own presence, and he made himself known by speaking as though to a friend (Exod 33:11), declaring himself apophatically as "he who is" (Exod 3:14). He appeared to them as "the personal hypostasis of God which is the comprehensive and exhaustive expression of His being," and this leads us to the fulness of his revelation within the church:

> The one God is not one divine nature or essence, but primarily one person: the person of God the Father. The personal existence of God (the Father) constitutes his essence of being, making it into "hypostases": freely and from love He begets the Son and causes the Holy Spirit to proceed. Consequently, being stems not from the essence, which would make it an ontological necessity, but from the person and the freedom of its love which "hypostasizes" being into a personal and trinitarian communion. God the Father's *mode of being* constitutes existence and life as a fact of love and personal communion.[94]

There is not space here to do more than gesture toward this personal Source of all things Who is beyond names, but this personal Logos must be praised as the Divine one "according to which each thing receives its existence." This is the source and being of the cosmos and the root characterization of named *persons* beyond faceless *individuals*. Moreover, the personal knowledge that comes from our personal Source bequeaths an ethical relationship of both knowing (*nous*) and living in relationship:

> Thus the spiritual intelligence knows all things through knowing the ideas or energies of which they are the expression. It is capable of a direct intuition of the inner and spiritual nature of everything that is, of which the visible form is but the outward mode, the apparent reality. Its way of knowing is through direct experience, and not by means of concepts or discursive reasoning.[95]

A core implication of this moral epistemology is that ethics is restored from hypothetical individuals—more akin to numbers within utilitarian calculations—to persons knowing and known by God. Ethics is the way

93. Yannaras, *Relational Ontology*, 16 (emphasis original).

94. Yannaras, *Relational Ontology*, 17–18 (emphasis original).

95. Sherrard, *Human Image, World Image*, 92.

of virtue for spiritually attuned persons who do not wish to stop within the sphere of mere academic theorizing. Hence though we risk more theological particularity than Wittgenstein seemingly ever accepted, his words still ring true: "God grant the philosopher insight into what lies in front of everyone's eyes" (*CV*, 63e). Or, similarly: "It's only by thinking even more crazily than philosophers do that you can solve their problems" (*CV*, 75e). In this context, to think more crazily is to call ethics away from its modernist theoretical underpinnings back to the cosmological song where divine reality sees man at the head of creation, but not "head" in the sense of *separated from* any more than the human head is separable from the body. Cosmology restores the unity of individual and world so that ethical virtue becomes the common pursuit of mankind embodied within the world. Ethics is cosmological precisely when persons are seen existing within a holistic universe that is knowable and known.

Cosmological ethics surpasses scientific theorizing ethics not by contradicting everything about theorizing but by recontextualizing the whole endeavor of ethics. The modern ethical mode of knowing, as Berdyaev notes, has placed a cleavage between reality and knowledge. "Knowledge is not *something*, but is *about something*."[96] Hence philosophy for Berdyaev inherently requires a relationship between man and the world, anthropology and cosmology, and this makes explicit the theological move within modern scientific theorizing and naturalism: "philosophy sees the world from the point of view of man, while science sees the world apart from man . . . [n]aturalism also sees the world from the point of view of man, but it does so secretly and will not admit it."[97] Philosophy for Berdyaev and Sherrard assumes that the cosmos is part of man, not man a part of the cosmos, and this link places morality rightly within the cosmos that is not a set of naturalized things but rather spiritual realities coinhering within the anthropological cosmos.[98] Without a prioritized hierarchy of anthropology, the cosmos has no meaning:

96. Berdyaev, *Destiny of Man*, 2 (emphasis original).

97. Berdyaev, *Destiny of Man*, 7.

98. To be clear, the union of anthropology and cosmology—and hierarchy of anthropology—presented in Sherrard and Berdyaev should in no way be connected with the movements that align biblical religion and creationism with destruction of the environment. Refuting this, Norman Wirzba's *The Paradise of God: Renewing Religion in an Ecological Age* is helpful for showing the heightened ecological responsibility of mankind at the head of creation rather than mankind's ownership or possession of creation. For instance, Wirzba points to the ethical calculus held by utilitarians such as Bentham and Singer who do not grant moral value to "sentient" beings that feel no pain. Beyond sentient life, however, is sentiotic life. When sentiosis as "the presence of a deep telos or proclivity towards created life" is primary, "then habitats and the organisms they support become candidates for moral considerability" (Wirzba, *Paradise of God*, 109).

> The meaning of things is revealed not through their entering into man who is passive in relation to them, but through man's creative activity reaching out to meaning beyond an unmeaning world. There is no meaning in the solid, concrete world of objects. Meaning is only revealed in man's activity and implies that reality is conformable to man. . . . Man's creativeness does not mean that he construes a world of his own. Meaning is to be found not in the object which enters the mind and not in the subject who construes a mental world, but in the spiritual world which is neither objective nor subjective but is pure activity and spiritual dynamics. The process of knowledge is a real event, and meaning is revealed in it actively, i.e., the darkness of being becomes light. Knowledge is spiritual life and a part of the reality which is being known.[99]

Berdyaev's philosophy entails a necessary harmony and coinherence of anthropology and cosmology as the narrative of mankind, man's creativeness, reveals meaning. Without mankind, there is no meaning, and yet the entire trajectory of modernist, scientific ethics removes mankind from the cosmos to seek ethical knowledge precisely where there is no meaning. Hence Berdyaev's philosophy inherently is ethical and tied to man as a being seeking salvation. "Ethics cannot be merely a theoretical philosophical discipline, for it is also a moral and spiritual activity."[100]

It may seem here that we are immersed in metaethics and far from bioethics, but this is not the case. *Bios* and the body have everything to do with cosmological ethics when we understand the narrative progression of the body in its perfect Edenic nature, its fallen imperfected nature, and then the body restored, transfigured, and deified by the incarnation. Symbolically, these modes of bodily life align with the architecture of the traditional cathedral where the altar represents the spirit of man, the sanctuary aligns with man's soul—namely his rational, willful, and appetitive powers—and the nave to man's body. Gleaning from St. Maximus's *Mystagogia*, Larchet draws out the ethical import of the body. Human beings symbolize the church

> not simply by their nature but by their own specific functions in spiritual life: the body represents in particular the practical or ethical dimension; the soul stands for the contemplative dimension; the spirit is its pinnacle, *theologia*, in which the believer receives from the Holy Spirit supernatural knowledge of the divine mysteries.[101]

99. Berdyaev, *Destiny of Man*, 8–9.
100. Berdyaev, *Destiny of Man*, 15.
101. Larchet, *Theology of the Body*, 84.

To further this symbolism in terms of temple theology, we may align day one in the Genesis account with the holy of holies, the second day to the veil, the third day to the bronze laver (or shewbread), and the fourth day to the seven-branched lamp—hence moving "down" the days of creation in parallel descent from the holy of holies (altar) past the curtain to the holy place (nave), then vestibule (narthex), etc. Margaret Barker sees the holy of holies as the throne of God, the hidden source of life and creation that emerges from the waters of chaos as light is divided from darkness in day one. In day two, the waters above and waters below are divided by the firmament heaven as a "web of matter [the Temple veil] that conceals the throne of God from human perception."[102] Day three shows forth the dry land Earth that is fruitful; the shewbread is upon the table with wine and incense. On day four, the lights in the heavens appear like the seven-branched lamp, while on day five, the altar of burning offering may be aligned with nonhuman creatures, birds flying aloft in the heavenly firmament and whales amidst the deeps. The sixth day accords with the high priest, mankind as caretaking priest of creation.[103]

This liturgical and ecclesial cosmological anthropology leads to a renewed ethics of wholeness where *bios* and the body are inherently icons of life and light. Modern scientific theorizing of the abstract *thing* or materialistic *object* is an iconoclasm, a paradigm that reshapes our mental outlook around the limited viewpoint to which the modern world clings. Science is a mirror, showing only the ego-face of the individual; sacred cosmology is an icon, revealing the higher meaning coinhering within the lower matter. For Sherrard, men such as Kepler, Galileo, Descartes, and Newton—differences notwithstanding—replaced ancient cosmology (meaning) with the method of mathematics (materialism). "They were themselves great mathematicians; and hence in default of any more profound understanding of things they identified their God with the great cosmic mathematician."[104] But the mind-soul is not a calculating *thing*, nor is the earth or body an external *thing*. The church fathers continually stress the coexistence and compound nature of mankind as ensouled body and embodied soul made in "our image" (*eikon*) and according to "our likeness" (Gen 1:26). "The image is actual, already realized; the likeness, on the other hand, is potential or virtual, something still to be accomplished."[105] The image of God remains a "permanent characteristic" of our human being, yet the likeness remains

102. Barker, *Temple Theology*, 18, 13–32. See also Pageau, *Language of Creation*.

103. Barker, *Temple Theology*, 18.

104. Sherrard, *Human Image, World Image*, 35.

105. Larchet, *Theology of the Body*, 27.

as our potential, our path of virtue that takes our human bodies from the limitations of fallenness into the restored fullness of the original Adamic body, the bodies of the saints, and ultimately Christ's body.

In this way, without forgoing the soul and spirit, the body transfigured and deified essentially is the ascetic path of ethics. The body is the particular focus of bioethics, and the way of virtue participating in Christ the man (anthropos) via bioethics places us in the nave while also indicating the directionality of our eschatological *telos*. Our progression should be past the nave to the narthex to the altar, and yet the progression "further up and further in" does not leave what is behind any more than *theologia* departs from the virtues. The pathway is dynamic such that the body, the soul, and the spirit coinhere. This is the mystical way of recontextualizing ethics away from calculations of utility and thin theories toward a reimagined health grounded on the reality of icons, saints, and the incarnation. With this as our ethical touchstone, the body transfigured imbues grace as the relics of saints defy corruption, or as St. Peter walks upon water, or as St. Mary of Egypt survives forty-seven years in the wilderness with only three loaves of bread, or as Christ bilocates (I Cor 5:3–8). The true nature of the body is not limited to the frail, fallen body within ancestral sin; the Edenic and resurrected body is normative.[106] In harmony with this, for any possible recontextualization of ethics, we must first move away from scientific ethics toward cosmological ethos; in other words, we begin properly "to relate directly temporal events and phenomena to their trans-temporal sources."[107] In short, the body is restored to "the place in the spiritual life that it should have" as coinhering matter, soul, and spirit. The *metania*, both small and great, is a comic act.[108]

A final implication of this is that we begin to see bioethics as something to be for and against; traditional Christians can accept only a reconfigured bioethics that moves past the scientific theorizing Wittgenstein so ardently critiques to embrace the cosmological significance of bodily matter and

106. "Those who live in intimacy with God are partly able to return to the condition of paradise: impassability becomes attainable again in this world; incorruptibility and immortality are promised as future realities. The lives of the saints show us in many ways that, when permeated with the divine energies, the body not only revels its destiny—one that transcends the normal lot of matter—but to some extent is already exempt from the laws of nature, as is demonstrated by such diverse extraordinary phenomena such as levitation, walking on water, the ability to see far-off things, bilocation, or surviving on no food other than Holy Communion. None of these manifestations has any value in itself, but they are signs that the other world, the kingdom of heaven, is a reality of which those who are worthy are granted an earnest here below, not only in their souls but their bodies" (Larchet, *Theology of the Body*, 88).

107. Sherrard, *Christianity*, 67.

108. Larchet, *Theology of the Body*, 82.

actions conducted in light of the eternal meaning. Hence the broader significance of day one in Genesis shows us the creation of heaven as meaning and earth as matter; where the transcendent sphere of heaven guides us, our embodied participation in God's divine law-nature *is* our ethic.

As the language of symbolism unites the spiritual, unified reality of the heavens with the diversity of matter and the earth, so this is where ethics, too, is symbolism. Ethics occurs at the meeting place of the one and the many, unity and diversity, heavens and the earth. As the heavens place their imprint upon matter and the body, the ethical life of mankind occurs within that meeting place. By taking embodied matter into the sphere of spiritual meaning, and by drawing meaning into and within our bodies, the work of ethics finds life-affirming traction and roots. Ethics becomes nothing less than the way of *theosis*, the destiny of mankind. As language unites the spheres of intangible thought and tangible word-grammar, as heaven and earth are united by the Word, so ethics unites microcosmic mankind as words of the Word.

Bibliography

Abelard, Peter. "Dialogue 2: Between the Philosopher and the Christian." In *Ethical Writings*, translated by Paul Vincent Spade, 91–148. Indianapolis: Hackett, 1995.

Adorno, Theodor. *Negative Dialectics*. Translated by E. B. Ashton. New York: Continuum, 1976.

Agamben, Giorgio. *The Highest Poverty: Monastic Rules and Form-of-Life*. Translated by Adam Kotsko. Stanford, CA: Stanford University Press, 2013.

Agar, Nicholas. *Humanity's End: Why We Should Reject Radical Enhancement*. Cambridge: MIT Press, 2010.

American Humanist Association. "Humanist Manifesto II." https://americanhumanist.org/what-is-humanism/manifesto2/.

American Society for Bioethics and Humanities. "Certification for Health Care Ethics Consultants: An Update." https://asbh.org/uploads/CECA_Announcement.pdf.

Anscombe, G. E. M. "Modern Moral Philosophy." *The Journal of the Royal Institute of Philosophy* 33 (1958) 1–19.

Anselm. *Monologion and Proslogion*. Translated by Thomas Williams. Indianapolis: Hackett, 1996.

Aquinas, Thomas. *Introduction to St. Thomas Aquinas*. Edited by Anton A. Pegis. New York: Modern Library, 1965.

Aristotle. *Metaphysics*. Translated by Hugh Lawson-Tancred. New York: Penguin Classics, 1999.

———. *Nichomachean Ethics*. Translated by Joseph Sachs. Newburyport, MA: Focus, 2002.

———. *Parts of Animals*. Translated by Arthur Leslie Peck and Edward Seymour Forster. Cambridge: Harvard University Press, 1937.

———. *Politics*. Translated by Benjamin Jowett. New York: Dover, 2000.

Arras, John. "The Hedgehog and the Borg: Common Morality in Bioethics." *Theoretical Medicine and Bioethics* 30 (2009) 11–30.

Asad, Talal. "Thinking about Tradition, Religion, and Politics in Egypt Today" *Critical Inquiry* 42 (2015) 166–214.

Athanasius. *On the Incarnation*. Translated by John Behr. Yonkers, NY: St. Vladimir's Seminary Press, 2011.

Augustine. *Confessions*. 2nd ed. Translated by Francis Joseph Sheed. Indianapolis: Hackett, 2006.

Bacon, Roger. *Novum Organum*. Translated and edited by Peter Urbach and John Gibson. Chicago: Open Court, 1994.

Barker, Margaret. *Temple Theology: An Introduction*. London: SPCK, 2004.

Bard, Bernard, and Joseph Fletcher. "The Right to Die." *The Atlantic Monthly* 221 (1968) 59–64.

Barrett, Cyril. *Wittgenstein on Ethics and Religious Belief*. Oxford: Blackwell, 1991.

Bartz, Robert. "Remembering the Hippocratics." In *Bioethics: Ancient Themes in Contemporary Issues*, edited by Mark Kuczewski and Roman Polansky, 3–30. Cambridge: MIT Press, 2000.

Beauchamp, Tom, and Leroy Walters. *Contemporary Issues in Bioethics*. Encino, CA: Dickenson, 1978.

———. *Principles of Biomedical Ethics*. 6th ed. Oxford: Oxford University Press, 2008.

———. *Principles of Biomedical Ethics*. 8th ed. Oxford: Oxford University Press, 2019.

Bentham, Jeremy. *The Principles of Morals and Legislation*. Amherst, NY: Prometheus, 1988.

Berdyaev, Nicolas. *The Destiny of Man*. Translated by Natalie Duddington. San Rafael, CA: Semantron, 2009.

Bess, Michael. "Enhanced Humans versus 'Normal People': Elusive Definitions." *Journal of Medicine and Philosophy* 35.6 (2010) 641–55.

The Bioethics Commision. "Gray Matters: Integrative Approaches for Neuroscience, Ethics and Society." Washington, DC: Presidential Commission for the Study of Bioethical Issues, May, 2014.

Bishop, Jeff. *The Anticipatory Corpse: Medicine, Power, and the Care of the Dying*. Notre Dame: University of Notre Dame Press, 2011.

———. "False Gods and Facades of the Same: On the Distinctiveness of a Christian Bioethics." *Christian Bioethics* 20.2 (2014) 301–17.

———. "Transhumanism, Metaphysics, and the Posthuman God." *Journal of Medicine and Philosophy* 35.6 (2010) 700–720.

Bishop, Jeff, and Fabrice Jotterand. "Bioethics as Biopolitics." *Journal of Medicine and Philosophy* 31.3 (2006) 205–12.

Borlaug, Norman. "The Green Revolution Revisited and the Road Ahead." https://www.nobelprize.org/prizes/peace/1970/borlaug/article/.

Bostrom, Nick. "Human Genetic Enhancements: A Transhumanist Perspective." *Journal of Value Inquiry* 37 (2003) 493–506.

———. "The Transhumanist FAQ, v. 2.1." Oxford: World Transhumanist Association, 2003.

Bouyer, Louis. *Cosmos: The World and the Glory of God*. Petersham, MA: St. Bede's, 1988.

Bradley, Denis. *Aquinas on the Twofold Human Good*. Washington DC: Catholic University Press, 1997.

Brody, Baruch. Lifetime Achievement Award Ceremony. Houston, TX: ASBH Annual Meeting, 2015.

Buchanan, Allen. "Social Moral Epistemology and Bioethicists." In *The Ethics of Bioethics: Mapping the Moral Landscape*, edited by Lisa Eckenwiler and Felicia Cohn, 288–96. Baltimore: The Johns Hopkins University Press, 2007.

Buchanan, James M., and Gordon Tullock. *The Calculus of Consent: Logical Foundations of Constitutional Democracy.* The Collected Works of James M. Buchanan 3. 20 vols. Indianapolis: Liberty Fund, 1999.

Burdett, Michael. "Transcendence and Human Enhancement: Francis Bacon, N. F. Fedorov, and Pierre Teilhard de Chardin." In *Transhumanism and Transcendence: Christian Hope in an Age of Technological Enhancement*, edited by Ronald Cole-Turner, 19–35. Washington DC: Georgetown University Press, 2011.

Callahan, Daniel. "Bioethics as a Discipline." *The Hastings Center Studies* 1.1 (1973) 66–73.

Carlisle, Clare, and Mark Sinclair. "Introduction and Commentary." In *Of Habit*, by Félix Ravaisson, 1–21. Translated by Clare Carlisle and Mark Sinclair. New York: Continuum International, 2008.

Carlton, Clark. "Wittgenstein, Apophasis, and Therapeutic Critique." Presentation, Society of Orthodox Philosophers in America, Holy Archangel's Monastery, Kendalia, TX, February 2012.

The Catholic Church. *Donum Vitae—Instruction on Respect for Human Life in Its Origin and on the Dignity of Procreation: Replies to Certain Questions of the Day.* Washington, DC: Office of Pub. and Promotion Services, United States Catholic Conference, 1987. https://www.vatican.va/roman_curia/congregations/cfaith/documents/rc_con_cfaith_doc_19870222_respect-for-human-life_en.html.

Chambers, Tod. "The Virtue of Attacking the Bioethicist." In *The Ethics of Bioethics: Mapping the Moral Landscape*, edited by Lisa Eckenwiler and Felicia Cohn, 281–87. Baltimore: The Johns Hopkins University Press, 2007.

Chardin, Pierre Teilhard de. *Activation of Energy: Enlightening Reflections on Spiritual Energy*. Translated by René Hague. London: Harcourt, 1978.

Cherry, Mark. "The Emptiness of Postmodern, Post-Christian Bioethics: An Engelhardtian Reevaluation of the Status of the Field." *Christian Bioethics* 20.2 (2014) 168–86.

Clack, Brian. *Introduction to Wittgenstein's Philosophy of Religion.* Edinburgh: Edinburgh University Press, 1999.

———. *Wittgenstein, Frazer and Religion*. London: MacMillan, 1999.

Cunningham, Connor. "Wittgenstein after Theology." In *Radical Orthodoxy: A New Theology*, edited by John Milbank et al., 64–90. London: Routledge, 1999.

Curran, Charles. "Curran: Papacy Should Admit Some of Its Teachings Are Wrong." *National Catholic Reporter*, November 21, 2014. https://www.ncronline.org/news/accountability/curran-papacy-should-admit-some-its-teachings-are-wrong.

Descartes, René. *Discourse on Method and Meditations on First Philosophy.* 4th ed. Translated by Donald A. Cress. Cambridge: Hackett, 1999.

Diamond, Cora. *The Realistic Spirit: Wittgenstein, Philosophy, and the Mind.* Cambridge: MIT Press, 1991.

Dostoevsky, Feodor. *The Brothers Karamazov*. Translated by Richard Pevear and Larissa Volokhonsky. New York: Farrar, Straus and Giroux, 2002.

Dummet, Michael A. E. *Truth and Other Enigmas*. London: Duckworth, 1978.

Dupré, Luis. *Passage to Modernity: An Essay in the Hermeneutics of Nature and Culture.* New Haven: Yale University Press, 1993.

Eckenwiler, Lisa, and Felicia Cohn, eds. *The Ethics of Bioethics: Mapping the Moral Landscape*. Baltimore: The Johns Hopkins University Press, 2007.

Edwards, James C. *Ethics without Philosophy: Wittgenstein and the Moral Life*. Tampa: University Presses of Florida, 1982.

Eisler, Rudolf. *Leib und seele: Darstellung und Kritik der Neueren Theorien des Verhältnisses Zwischen Psychischem und Physischem Dasein*. Leipzig: Barth, 1906.

Elliott, Carl. "The Tyranny of Expertise." In *The Ethics of Bioethics: Mapping the Moral Landscape*, edited by Lisa Eckenwiler and Felicia Cohn, 43–46. Baltimore: The Johns Hopkins University Press, 2007.

Engelhardt, H. T. E. *After God: Morality and Bioethics in a Secular Age*. Yonkers, NY: St. Vladimir's Orthodox Seminary Press, 2017.

———. "Common Ground as Battleground: The Culture Wars Framing Bioethics Continue." *O Mundo da Saude* 39 (2015) 514–30.

———. "The Deprofessionalization of Medicine in the United States: From Guild to Managed Care." *City University of Hong Kong Occasional Paper Series 9* (2001) 1–13.

———. "Foreword." In *Global Bioethics: Building on the Leopold Legacy*, by Van Rensselaer Potter, vii–xii. East Lansing: Michigan State University Press, 1988.

———. *The Foundations of Bioethics*. 2nd ed. Oxford: Oxford University Press, 1996.

———. "The Foundations of Bioethics: Rethinking the Meaning of Morality." In *The Story of Bioethics: From Seminal Works to Contemporary Explorations*, edited by Jennifer K. Walter and Eran P. Klein, 91–112. Washington DC: Georgetown University Press, 2003.

———. "The Foundations of Bioethics and Secular Humanism: Why Is There No Canonical Moral Content?" In *Reading Engelhardt: Essays on the Thought of H. Tristram Engelhardt, Jr.*, edited by Brendan P. Minogue et al., 259–86. Dordrecht: Kluwer Academic, 1997.

———. *The Foundations of Christian Bioethics*. Lisse: Swets and Zeitlinger, 2000.

———. "Philosophy of Medicine." In *A Guide to the Culture of Science, Technology, and Medicine*, edited by Paul T. Durbin, 364–461. New York: Free Press, 1980.

———. "The Recent History of Christian Bioethics Critically Reassessed." *Christian Bioethics* 20.2 (2014) 146–67.

———. "Re-Reading *Re-Reading Engelhardt*." In *At the Roots of Christian Bioethics: Critical Essays on the Thought of H. Tristram Engelhardt, Jr.*, edited by Ana S. Iltis and Mark J. Cherry, 285–316. Salem, MA: Scrivener, 2010.

Engelmann, Paul. *Letters from Ludwig Wittgenstein: With a Memoir*. New York: Horizon, 1967.

Evagrius. *Evagrius Ponticus*. Edited by Augustine Casiday. Oxford: Routledge, 2006.

———. "On Prayer: One Hundred and Fifty-Three Texts." In *The Philokalia*, edited by St. Nikokimos of the Holy Mountain & St. Makarios of Corinth, translated by Gerald E. H. Palmer et al., 1:55–72. 4 vols. New York: Faber & Faber, 1983.

Evans, John. *The History and Future of Bioethics: A Sociological View*. Oxford: Oxford University Press, 2012.

———. *Playing God? Human Genetic Engineering and the Rationalization of Public Bioethical Debate*. Chicago: The University of Chicago Press, 2002.

Fan, Ruping. ed. *Confucian Bioethics*. Dordrecht: Kluwer, 2012.

———. "The Unjustifiability of Substantive Liberalisms and the Inevitability of Engelhardtian Procedural Liberalism." In *Reading Engelhardt: Essays on the*

Thought of H. Tristram Engelhardt, edited by Brendan P. Minogue et al., 221–36. Dordrecht: Kluwer Academic, 1997.

Fechner, Gustav Theodor. *Elemente der Psychophysik*. Leipzig: Drück und Verlag von Breitkopf und Hartel, 1860.

———. *Nanna: Über das Seelenleben der Pflanzen*. Leipzig: Leopold Voß, 1848.

Fletcher, Joseph. *Medicine and Morals*. Boston: Beacon, 1960.

Fogelin, Robert. "Wittgenstein's Critique of Philosophy." In *The Cambridge Companion to Wittgenstein*, edited by Hans Sluga and David G. Stern, 34–58. Cambridge: Cambridge University Press, 1996.

Folz, Bruce. *The Noetics of Nature: Environmental Philosophy and the Holy Beauty of the Visible*. New York: Fordham University Press, 2014.

Foster, Charles. *Choosing Life, Choosing Death: The Tyranny of Autonomy and Medical Ethics and Law*. Oxford: Hart, 2009.

Foucault, M. *The Birth of the Clinic*. New York: Vintage, 1994.

———. *Madness and Civilization*. New York: Random House, 1988.

———. "Right of Death and Power Over Life." In *The Foucault Reader*, edited by Paul Rabinow, 258–72. New York: Vintage, 2010.

Fukuyama, Francis. "Transhumanism." *Foreign Policy*, October 23, 2009. http://foreignpolicy.com/2009/10/23/transhumanism/.

Galen. *Method of Medicine, Vol. I, Books 1–4*. 1st ed. 3 vols. Translated by Ian Johnston and Greg H. R. Horsley. Cambridge: Loeb Classical Library, 2011.

Garner, Stephen. "The Hopeful Cyborg." In *Transhumanism and Transcendence: Christian Hope in an Age of Technological Enhancement*, edited by Ronald Cole-Turner, 87–100. Washington DC: Georgetown University Press, 2011.

Garver, Newton. "Philosophy as Grammar." In *The Cambridge Companion to Wittgenstein*, edited by Hans Sluga and David G. Stern, 139–70. Cambridge: Cambridge University Press, 1996.

Goss, Boaz, and Rico Vitz. "Natural Law among Moral Strangers." *Christian Bioethics* 20.2 (2014) 283–300.

Grant, Robert M. and Holt. H. Graham, eds. *Apostolic Fathers, Vol. II*. 6 vols. Nashville: Nelson and Sons, 1965.

Gregory of Nyssa. *Life of Moses*. Translated by Abraham J. Malherbe and Everett Ferguson. Mahwah, NJ: Paulist, 1978.

Gustafson, James. "Theology Confronts Technology and the Life Sciences." *Commonweal* 105 (1978) 386–92.

Habermas, Jürgen. *Theory and Practice*. Translated by John Viertel. Boston: Beacon, 1973.

Hacker, Peter M. S. *Wittgenstein: Comparisons and Context*. Oxford: Oxford University Press, 2013.

———. *Wittgenstein: Connections and Controversies*. Oxford: Clarendon, 2001.

Hadot, Pierre. *Philosophy as a Way of Life*. Edited by A. Davidson. Oxford: Blackwell, 1995.

———. "Wittgenstein, philosophe du langage (II)." *Critique* 150 (November 1959) 972–83.

Hartmann, Klaus. "On Taking the Transcendental Turn." *Review of Metaphysics* 67 (1966) 224–25.

Hauerwas, Stanley. "How Christian Ethics Became Medical Ethics: The Case of Paul Ramsey." *Christian Bioethics* 1.1 (1995) 11–28.

———. "Not All Peace Is Peace." In *Reading Engelhardt: Essays on the Thought of H. Tristram Engelhardt*, edited by Brendan P. Minogue et al., 31–44. Dordrecht: Kluwer Academic, 1997.

———. "Theological Reflections on In Vitro Fertilization." In *HEW Support of Research Involving Human In Vitro Fertilization and Embryo Transfer*, 1–20. Washington DC: US Government Printing Office, 1979. https://archive.org/stream/hewsupportofrese02unit/hewsupportofrese02unit_djvu.txt.

Hauskeller, Michael. "Nietzsche, the Overhuman and the Posthuman: A Reply to Stefan Sorgner." *Journal of Evolution and Technology* 21.1 (January 2010) 5–8.

Hegel, G. W. F. *Elements of the Philosophy of Right or Natural Law and Political Science in Outline*. Edited by Allen W. Wood. Translated by Hugh B. Nisbet. Cambridge: Cambridge University Press, 2003.

———. *Phenomenology of Spirit*. Translated by Arnold Miller. Oxford: Oxford University Press, 1977.

Heidegger, Martin. "Letter on Humanism." In *Martin Heidegger: Basic Writings*, edited and translated by David Farrell Krell, 213–66. New York: Harper Collins, 1993.

Hippocrates. *Hippocratic Writings*. Edited by Geoffrey E. R. Lloyde et al. Translated by John Chadwick et al. New York: Penguin Classics, 1984.

Hobbes, Thomas. *Leviathan*. Edited by Edwin Curley. Indianapolis: Hackett, 1994.

Horwich, Paul. *Wittgenstein's Metaphilosophy*. Oxford: Oxford University Press, 2012.

Hughes, James. "Contradictions from the Enlightenment Roots of Transhumanism." *Journal of Medicine and Philosophy* 35.6 (2010) 622–40.

———. "Morality in a Pill?" *Institute for Ethics and Emerging Technology*, October 9, 2012. https://archive.ieet.org/articles/hughe20121009.html.

———. "Problems of Transhumanism: The Unsustainable Autonomy of Reason." *Institute for Ethics and Emerging Technology*, January 8, 2010. https://archive.ieet.org/articles/hughes20100108.html.

Incandela, Joseph. "An Augustinian Reading of Aquinas and Wittgenstein." In *Grammar and Grace: Reformulations of Aquinas and Wittgenstein*, edited by Jeffery Stout and Robert MacSwain, 24–50. London: SCM, 2004.

Ip, Pu Him. "Identifying the 'Apophatic Impulse' in Ludwig Wittgenstein's Early Philosophy: The Lecture on Ethics as an Interpretive Key." In *Ludwig Wittgenstein between Analytic Philosophy and Apophaticism*, edited by Sotiris Mitralexis, 81–106. Cambridge: Cambridge Scholars, 2015.

Isaac the Syrian. *The Ascetical Homilies*. Brookline, MA: Holy Transfiguration Monastery, 2011.

Jahr, Fritz. "Bio-Ethik. Eine Umschau über die ethischen Beziehungen des Menschen zu. Tier und Pflanze." *Kosmos: Handweiser fur Naturfreunde* 24 (1927) 2–4.

Janik, Allan. "Wittgenstein's Strategy." *The New York Review*, July 10, 1969. https://www.nybooks.com/articles/1969/07/10/wittgensteins-strategy-2/.

Johnson, Mark. *Moral Imagination: Implications of Cognitive Science for Ethics*. Chicago: University of Chicago Press, 1993.

Jones, David Albert. "An Unholy Mess: Why 'The Sanctity of Life Principle' Should Be Jettisoned." *The New Bioethics* 22:3 (2016) 185–201.

Jonsen, Albert. *The Birth of Bioethics*. Oxford: Oxford University Press, 1998.

———. "Lifetime Achievement Award Ceremony." ASBH Annual Meeting, Philadelphia, Pennsylvania, October 29, 1999.

———. "Why Has Bioethics Become so Boring?" *Journal of Medicine and Philosophy*, 25.6 (2000) 689–99.

Jonsen, Albert, et al. *Clinical Ethics*. 7th ed. New York: McGraw-Hill Education, 2010.

Jonston, Paul. *The Contradictions of Modern Moral Philosophy: Ethics after Wittgenstein*. Indianapolis: Routledge, 1999.

Jotterand, Fabrice. "At the Roots of Transhumanism: From the Enlightenment to a Post-Human Future." *Journal of Medicine and Philosophy* 35 (2010) 617–21.

Juvin, Hervé. *The Coming of the Body*. Translated by John Howe. London: Verso, 2010.

Kallenberg, Brad. *Ethics as Grammar: Changing the Post-Modern Subject*. Notre Dame: University of Notre Dame Press, 2001.

Kant, Immanuel. *Critique of Pure Reason*. Translated by Norman Kemp Smith. New York: St. Martin's, 1965.

———. *Ethical Philosophy: Grounding for the Metaphysics of Morals, Metaphysical Principles of Virtue, with "On a Supposed Right to Lie Because of Philanthropic Concerns."* 2nd ed. Translated by James W. Ellington. Indianapolis: Hackett, 1995.

———. *Grounding for the Metaphysics of Morals*. 3rd ed. Translated by. James W. Ellington. Indianapolis: Hackett, 1993.

Kass, Leon. *Life, Liberty and the Defense of Dignity: The Challenge for Bioethics*. San Francisco: Encounter, 2002.

———. "The New Genetics and the Future of Man." In *New Beginnings in Life*, edited by Michael P. Hamilton, 13–63. Grand Rapids: Eerdmans, 1972.

———. "The Wisdom of Repugnance." *The New Republic* (June 2, 1997) 17–26.

Katongole, Emmanuel. *Beyond Universal Reason: The Relation Between Religion and Ethics in the Work of Stanley Hauerwas*. Notre Dame: University of Notre Dame Press, 2000.

Keown John, and David Jones. "Surveying the Foundations of Medical Law: A Reassessment of Glanville Williams's *The Sanctity of Life and the Criminal Law*." *Medical Law Review* 16.1 (2008) 85–126.

Kerr, Fergus. "Metaphysics and Magic: Wittgenstein's Kink." In *Post-Secular Philosophy: Between Philosophy and Theology*, edited by Phillip Blond, 240–58. London: Routledge, 1998.

———. *Theology after Wittgenstein*. 2nd ed. London: Society for Promoting Christian Knowledge, 1997.

Kierkegaard, Søren. *Concluding Unscientific Postscript to Philosophical Fragments*. Edited and translated by Howard and Edna Kong. Princeton: Princeton University Press, 1992.

———. *The Point of View*. Edited and translated by Howard and Edna Kong. Princeton: Princeton University Press, 1998.

Klugman, Craig. "Bioethics: The Revolution Is Over." *Bioethics Today*, November 10, 2015. http://www.bioethics.net/2015/11/bioethics-the-revolution-is-over/.

Koch, Tom. *Thieves of Virtue: When Bioethics Stole Medicine*. Cambridge: MIT Press, 2012.

Kopelman, Loretta. "The Growth of Bioethics as a Second-Order Discipline." In *The Development of Bioethics in the United States*, edited by Jeremy R. Garrett et al., 137–59. Dordrecht: Springer, 2012.

Kuhn, Thomas. *The Structure of Scientific Revolutions*. 4th ed. Chicago: University of Chicago Press, 2012.

Kymlicka, Will. *Contemporary Political Philosophy: An Introduction*. 2nd ed. Oxford: Oxford University Press, 2001.

Labron, Tim. *Wittgenstein and Theology*. London: T. & T. Clark, 2009.

Lacoste, Jean-Yves. *From Theology to Theological Thinking*. Translated by Christopher Hackett. Charlottesville: University of Virginia Press, 2014.

Larchet, Jean-Claude. *Theology of the Body*. Translated by Michael Donley. Yonkers, NY: St. Vladimir's Orthodox Seminary Press, 2016.

Laugier, Sandra. "Pierre Hadot as a Reader of Wittgenstein." Translated by James Helgeson. *Paragraph* 34.3 (2011) 322–37.

Lee, Marvin. "The Problem of 'Thick in Status, Thin in Content' in Beauchamp and Childress' Principlism." *Journal of Medical Ethics* 36.9 (September 2010) 525–28.

Locke, John. *Second Treatise of Government*. Edited by Crawford B. Macpherson. Indianapolis: Hackett, 1980.

Long, D. Stephen. *Speaking of God: Theology, Language, and Truth*. Grand Rapids: Eerdmans, 2009.

Lossky, Vladimir. *The Mystical Theology of the Eastern Church*. Yonkers, NY: St. Vladimir's Orthodox Seminary Press, 1976.

Lucie-Smith, Alexander. *Narrative Theology and Moral Theology: The Infinite Horizon*. Hampshire, UK: Ashgate, 2007.

Lysaught, M. Therese. "Docile Bodies: Transnational Research Ethics as Biopolitics." *The Journal of Medicine and Philosophy* 34.4 (2009) 384–408.

MacIntyre, Alasdair. *After Virtue*. 3rd ed. Notre Dame: University of Notre Dame Press, 2007.

———. *Whose Justice? Which Rationality?* Notre Dame: University of Notre Dame Press, 1988.

Maclean, Anne. *The Elimination of Morality: Reflections on Utilitarianism and Bioethics*. Indianapolis: Routledge, 1993.

Malcolm, Norman. *Wittgenstein: A Religious Point of View?* Ithaca, NY: Cornell University Press, 1994.

Maximus the Confessor. *Maximus Confessor: Selected Writings*. Edited by George Berthold. Mahwah, NJ: Paulist, 1985.

———. *On the Cosmic Mystery of Jesus Christ: St. Maximus the Confessor*. Translated by Paul M. Blower and Robert L. Wilken. Yonkers, NY: St. Vladimir's Orthodox Seminary Press, 2003.

———. *Various Texts on Theology, the Divine Economy, and Virtue and Vice*. In *The Philokalia, Vol. 2*, edited by St. Nikokimos of the Holy Mountain and St. Makarios of Corinth, translated by Gerald E. H. Palmer et al., 2:164–84. 4 vols. New York: Faber and Faber, 1984.

McCormick, Richard. *Notes on Moral Theology: 1965–1980*. Lanham, MD: University Press of America, 1980.

———. "Theology and Bioethics." *The Hastings Center Report* 19.2 (March–April 1989) 5–10.

McDermott, John. "The Collapse of the Manualist Tradition." *Faith*, January–February, 2014. https://www.faith.org.uk/article/january-february-2014-the-collapse-of-the-manualist-tradition.

McFadden, Charles. *Medical Ethics*. 5th ed. Philadelphia: Davis, 1961.

McGuckin, John. *Westminster Handbook to Patristic Theology*. Louisville: Westminster John Knox, 2004.

McKenny, Gerald. "Transcendence, Technological Enhancement, and Christian Theology." In *Transhumanism and Transcendence: Christian Hope in an Age of Technological Enhancement,* edited by Ronald Cole-Turner, 177–92. Washington DC: Georgetown University Press, 2011.

Meillassoux, Quintin. *After Finitude: An Essay on the Necessity of Contingency.* Translated by Ray Brassier. London: Continuum International Publishing Group, 2009.

Meyendorff, John. *A Study of Gregory Palamas.* Translated by George Lawrence. Crestwood, NY: St. Vladimir's Seminary Press, 1964.

Milbank, John. *Beyond Secular Order: The Representation of Being and the Representation of People.* Oxford: Wiley-Blackwell, 2014.

———. *Theology and Social Theory: Beyond Secular Reason 2nd ed.* Oxford: Wiley-Blackwell, 2006.

Mill, John Stuart. *Utilitarianism.* 2nd ed. Edited by George Sher. Indianapolis: Hackett, 2002.

Mitralexis, Sotiris. "An Apophatic Wittgenstein—Or a Wittgensteinian Apophaticism." In *Ludwig Wittgenstein between Analytic Philosophy and Apophaticism*, edited by Sotiris Mitralexis, vii–xii. Cambridge: Cambridge Scholars, 2015.

Monk, Ray. *How to Read Wittgenstein.* New York: Norton, 2005.

———. *Ludwig Wittgenstein: The Duty of Genius.* New York: Penguin, 1990.

Moyal-Sharrock, Danièle. *The Third Wittgenstein: The Post-Investigations Works.* Aldershot, UK: Ashgate, 2004.

Mumford, Lewis. *The Myth of the Machine.* London: Secker and Warburg, 1967.

———. *Techniques and Civilization.* New York: Harcourt, Brace, 1934.

Nash, Ryan. "H. T. Engelhardt, Jr., Misrepresented and Misunderstood: Beyond a Philosophy of Difference and Toward a Theology of Unity." *Christian Bioethics* 20.2 (2014) 272–82.

The National Commission for the Protection of Human Subjects of Biomedical and Behavioral Research. *The Belmont Report: Ethical Principles and Guidelines for the Protection of Human Subjects of Research.* Bethesda, MD: The National Commission for the Protection of Human Subjects of Biomedical and Behavioral Research, 1978.

Nietzsche, Friedrich. *Beyond Good and Evil.* In *Basic Writings of Nietzsche,* translated and edited by Walter Kauffman, 179–436. New York: Modern Library Edition, 1992.

Ockham, William. *Philosophical Writings.* Edited and translated by Philotheus Boehner, OFM. Cambridge: Hackett, 1990.

Pageau, Matthew. *The Language of Creation: Cosmic Symbolism in Genesis, a Commentary.* N.p.: Create Space, 2018.

Palamas, Gregory. *The Triads.* Translated by Nicholas Gendle. Edited by John Meyendorff. Mahwah, NJ: Paulist, 1983.

Paulson, Steve. "The Philosopher Who Says We Should Play God: Why Ethical Objections to Interfering with Nature Are Too Late." *Nautilus,* August 31, 2015. https://nautil.us/the-philosopher-who-says-we-should-play-god-rp-235608/.

Pellegrino, Edmund, and David Thomasma. "From Medical Ethics to a Moral Philosophy of the Professions." In *The Story of Bioethics: From Seminal Works to Contemporary Explorations,* edited by Jennifer K. Walter and Eran P. Klein, 3–15. Washington DC: Georgetown University Press, 2003.

———. *A Philosophical Basis of Medical Practice: Toward a Philosophy and Ethic of the Healing Professions*. Oxford: Oxford University Press, 1981.

Percival, Thomas. *Percival's Medical Ethics*. South Yarra, VIC: Leopold Classic Library, 2016.

Persson, Ingmar, and Julian Savulescu. "Moral Transhumanism." *Journal of Medicine and Philosophy* 35 (2010) 656–69.

———. "The Perils of Cognitive Enhancement and the Urgent Imperative to Enhance the Moral Character of Humanity." *Journal of Applied Philosophy* 25.3 (2008) 162–77.

———. *Unfit for the Future: The Need for Moral Enhancement*. Oxford: Oxford University Press, 2012.

Peterson, Jordan. *Maps of Meaning: The Architecture of Belief*. Indianapolis: Routledge, 1999.

Plato, *Complete Works*. Edited by John M. Cooper. Indianapolis: Hackett, 1997.

Plested, Marcus. "Wisdom in the Fathers: An (Eastern) Orthodox Perspective." In *Encounter between Eastern Orthodoxy and Radical Orthodoxy: Transfiguring the World through the Word*, edited by Adrian Pabst and Christoph Schneider, 239–48. Indianapolis: Routledge, 2009.

Polansky, Ronald. "Is Medicine Art, Science, or Practical Wisdom? Ancient and Contemporary Reflections." In *Bioethics: Ancient Themes in Contemporary Issues*, edited by Mark G. Kuczewski, and Ronald Polansky, 31–56. Cambridge: MIT Press, 2000.

Polanyi, Michael. *Knowing and Being: Essays by Michael Polanyi*. Chicago: University of Chicago Press, 1969.

———. *Personal Knowledge: Towards a Post-Critical Philosophy*. Chicago: University of Chicago Press, 1974.

———. "The Republic of Science." *Minerva* 38 (2000) 1–32.

Potter, Van Rensselaer. *Bioethics: Bridge to the Future*. Englewood Cliffs, NJ: Prentice-Hall, 1971.

———. "Bioethics, the Science of Survival." *Perspectives in Biology and Medicine* 14 (1970) 127–53.

———. *Global Bioethics: Building on the Leopold Legacy*. East Lansing: Michigan State University Press, 1988.

Preller, Victor. *Divine Science and the Science of God: A Reformulation of Thomas Aquinas*. Eugene, OR: Wipf and Stock, 2005.

Pseudo-Dionysius the Areopagite. *The Complete Works*. Translated by Colm Luibheid. New York: Paulist, 1987.

Ramsay, Paul. *Patient as Person: Explorations in Medical Ethics*. 2nd ed. New Haven: Yale University Press, 2002.

Ravaisson, Félix. *Of Habit*. Translated by Claire Carlisle and Mark Sinclair. New York: Continuum International, 2008.

Rawls, John. *A Theory of Justice*. 2nd ed. Harvard, MA: Belknap, 1999.

Reich, Warren. *Encyclopedia of Bioethics*. New York: Free Press, 1978.

———. "How Bioethics Got Its Name." *Hastings Center Report* 23.6 (November–December 1993) 6–7.

———. "The Word 'Bioethics': The Struggle over Its Earliest Meanings." *Kennedy Institute of Ethics Journal* 5.1 (1995) 19–34.

Reno, Russel. *The Ordinary Transformed: Karl Rahner and the Christian Vision of Transcendence*. Grand Rapids: Eerdmans, 1995.

Rhees, Rush, ed. *Ludwig Wittgenstein: Personal Recollections*. Oxford: Blackwell, 1981.

———. *Wittgenstein's* On Certainty*: There—Like Our Life*. Edited by Dewi Z. Phillips. Oxford: Blackwell, 2003.

Rhodes, Rosamond. "Rethinking of Research Ethics." *American Journal of Bioethics* 5 (2005) 7–28.

Rosen, George. *A History of Public Health*. New York: MD, 1958.

Roser, Max, et al. "World Population Growth." http://ourworldindata.org/world-population-growth/.

Rousseau, Jean-Jacques. *The Basic Political Writings*. 2nd ed. Translated and edited by Donald A. Cress. Cambridge: Hackett, 2012.

Russell, Bertrand. *History of Western Philosophy*. London: Routledge, 1996.

Sagan, Carl. "Why We Need to Understand Science." *The Skeptical Inquirer* 14.3 (Spring 1990) 263–69.

Saidel, Matias Leandro. "Form(s)-of-Life. Agamben's Reading of Wittgenstein and the Potential Uses of a Notion." *Trans/Form/Acao Marilia* 37:1 (January 2014) 163–86.

Sass, Hans-Martin. "Fritz Jahr's 1927 Concept of Bioethics." *Kennedy Institute of Ethics Journal* 17.4 (2007) 279–95.

Savickey, Beth. *Wittgenstein's Art of Investigation*. London: Routledge, 1999.

Savulescu, Julian. "Bioethics: Why Philosophy Is Essential for Progress." *Journal of Medical Ethics* 41.1 (2015) 28–33.

———. "Clone Me Up, Scotty: A Brief Satirical History of Cloning and Ethical Progress." *Practical Ethics,* August, 20, 2015. http://blog.practicalethics.ox.ac.uk/2015/08/clone-me-up-scotty-a-brief-history-of-cloning-and-ethical-progress/.

Schönbaumsfeld, Genia. *A Confusion of the Spheres: Kierkegaard and Wittgenstein on Philosophy and Religion*. Oxford: Oxford University Press, 2007.

Scotus, John Duns. *Philosophical Writings*. 2nd ed. Translated by Allan B. Wolter, OFM. Indianapolis: Hackett, 1987.

Sherrard, Philip. *Christianity: Lineaments of a Sacred Tradition*. Brookline, MA: Holy Cross Orthodox Press, 2007.

———. *Human Image, World Image: The Death and Resurrection of Sacred Cosmology*. Limni, Greece: Denise Harvey, 2004.

———. *The Rape of Man and Nature: An Inquiry into the Origins and Consequences of Modern Science*. Limni, Greece: Denise Harvey, 2020.

Sidiropoulou, Chryssi. "Wittgenstein's Apophatic Descriptions." In *Ludwig Wittgenstein: between Analytic Philosophy and Apophaticism*, edited by Sotiris Mitralexis, 49–80. Cambridge: Cambridge Scholars, 2015.

Singer, Peter. *Practical Ethics*. 3rd ed. Cambridge: Cambridge University Press, 2011.

Sinsheimer, Robert L. "The Prospect for Designed Genetic Change." *American Scientist* 57.1 (Spring 1969) 134–43.

Sivrides, George. "Wittgenstein and Kierkegaard's Ethics: Ethics as Apophatic Knowledge." In *Ludwig Wittgenstein: Between Analytic Philosophy and Apophaticism*, edited by Sotiris Mitralexis, 107–26. Cambridge: Cambridge Scholars, 2015.

Smith, Christian. *Moral, Believing Animals: Human Personhood and Culture*. Oxford: Oxford University Press, 2009.

Smith, James K. A. *Who's Afraid of Relativism? Community, Contingency, and Creaturehood*. Grand Rapids: Baker Academic, 2014.

Smith, Wesley J. "The Deprofessionalization of Medicine." *National Review*, January 18, 2016. http://www.nationalreview.com/human-exceptionalism/429938/deprofessionalization-medicine.

Solomon, David. "Christian Bioethics, Secular Bioethics, and the Claim to Cultural Authority." *Christian Bioethics* 11.3 (2005) 349–59.

Solzhenitsyn, Aleksandr. *The Gulag Archipelago, 1918–1956: An Experiment in Literary Investigation*. 1st ed. New York: Harper & Row, 1974.

Sontag, Frederick. *Wittgenstein and the Mystical: Philosophy as an Ascetic Practice*. Atlanta: Scholars, 1995.

Spielgelberg, Herbert. "The Puzzle of Ludwig Wittgenstein's *Phänomenologie* (1929-?)." *American Philosophical Quarterly* 5.4 (1968) 244–56.

Stout, Jeffery. *The Flight from Authority: Religion, Morality, and the Quest for Autonomy*. Notre Dame: University of Notre Dame Press, 1981.

Sulmasy, Daniel. "In Defense of the Amphibians: A Critical Appraisal of Engelhardt on the Recent History of Christian Bioethics." *Christian Bioethics* 20.2 (2014) 187–95.

Taylor, Charles. *The Ethics of Authenticity*. Cambridge: Harvard University Press, 1992.

———. *Modern Social Imaginaries*. Durham, NC: Duke University Press, 2004.

Thompson, Ian. *Heidegger on Ontotheology: Technology and the Politics of Education*. Cambridge: Cambridge University Press, 2005.

Tocqueville, Alexis de. *Democracy in America*. Translated by Arthur Goldhammer. New York: The Library of America, 2004.

Tolstoy, Leo. *The Gospel in Brief: The Life of Jesus*. Translated by Dustin Condren. New York: HarperCollins, 2011.

Toulmin, Stephen. *The Return to Cosmology: Postmodern Science and the Theology of Nature*. Oakland: University of California Press, 1985.

Trader, Alexis. *Ancient Christian Wisdom and Aaron Beck's Cognitive Therapy: A Meeting of Minds*. New York: Lang, 2011.

Trotsky, Leon. *Their Morals and Ours: The Marxist View of Morality*. NSW, Australia: Resistance, 2000.

Trotter, Griffin. "Bioethics and Deliberative Democracy: Five Warnings from Hobbes." *Journal of Medicine and Philosophy* 31.3 (2006) 235–50.

———. "Is 'Discursive Christian Bioethics an Oxymoron?" In *At the Roots of Christian Bioethics: Critical Essays on the Thought of H. Tristram Engelhardt, Jr.*, edited by Ana S. Iltis and Mark J. Cherry, 203–28. Salem, MA: Scrivener, 2010.

Tsai, George. "Lamentable Necessities." *The Review of Metaphysics* 66.4 (2013) 775–808.

Vaughn, Lewis. *Bioethics, Principles, Issues, and Cases*. 4th ed. Oxford: Oxford University Press, 2019.

Veach, Robert. "Revisiting a Theory of Medical Ethics: Main Themes and Anticipated Changes." In *The Story of Bioethics: From Seminal Works to Contemporary Explorations*, edited by Jennifer K. Walter and Eran P. Klein, 67–90. Washington DC: Georgetown University Press, 2003.

Verdi, John. "Doing Things with Words." *The College* (Winter 2010) 10–15.

Walters, LeRoy. *Bibliography of Bioethics, Vol. 35*. 35 vols. Washington DC: Georgetown University Press, 2009.

Waters, Brent. *From Human to Posthuman: Christian Theology and Technology in a Postmodern World*. Indianapolis: Routledge, 2006.

Weber, Max. *The Protestant Ethic and the "Spirit" of Capitalism*. New York: Penguin Classics, 2002.

Weil, Simone. *Waiting for God*. New York: Putnam's Son, 1951.

Weinberg, Alvin M. "Science and Trans-Science." *Minerva* 10.2 (1974) 209–22.

Whitney, Charles. "Francis Bacon's *Instauratio*: Dominion of and over Humanity." *Journal of the History of Ideas* 50.3 (1989) 371–90.

Wildes, Kevin Wm, SJ. "Completing the Picture: Engelhardt's Christian Bioethics." In *At the Roots of Christian Bioethics: Critical Essays on the Thought of H. Tristram Engelhardt, Jr.*, edited by Ana S. Iltis and Mark J. Cherry, 89–106. Salem, MA: Scrivener, 2010.

———. "Engelhardt's Communitarian Ethics: The Hidden Assumptions." In *Reading Engelhardt: Essays on the Thought of H. Tristram Engelhardt, Jr.*, edited by Brendan P. Minogue et al., 77–94. Dordrecht: Kluwer Academic, 1997.

William, Glanville. *The Sanctity of Life and the Criminal Law*. Cambridge: Cambridge University Press, 2013.

Williamson, Timothy. "Must Do Better." In *Truth and Realism*, edited by Patrick Greenough and Michael Lynch, 177–87. Oxford: Oxford University Press, 2005.

Wirzba, Norman. *The Paradise of God: Renewing Religion in an Ecological Age*. Oxford: Oxford University Press, 2003.

Wittgenstein, Ludwig. *The Blue and Brown Books*. Oxford: Blackwell, 1958.

———. *Culture and Value*. Edited by. Georg H. von Wright and Heikki Nyman. Translated by Peter Winch. Oxford: Blackwell, 1980.

———. "Open Access to Transcriptions of the Wittgenstein Nachlass (2016-)." *The Wittgenstein Archives at the University of Bergen*, 2016. http://wab.uib.no/wab_transcriptions.page.

———. *Lecture on Ethics*. Edited by Edoardo Zamuner et al. Oxford: Wiley Blackwell, 2014.

———. *Letters from Ludwig Wittgenstein with a Memoir by Paul Engelmann*. Edited by Brian F. McGuinness. Oxford: Blackwell, 1967.

———. "Letters to Ludwig von Ficker." In *Wittgenstein: Sources and Perspective*, edited by Allan Janik and C. Grant Luckhardt, translated by Bruce Gillette, 82–98. Cornell, NY: Cornell University Press, 1979.

———. *Letters to Russell, Keynes, and Moore*. Edited by Georg H. von Wright. Oxford: Blackwell, 1974.

———. *Ludwig Wittgenstein: Personal Recollections*. Edited by Rush Rhees. Oxford: Blackwell, 1981.

———. *Ludwig Wittgenstein and the Vienna Circle: Conversations Recorded by Friedrich Waismann*. Edited by Brian McGuinness. Translated by Joachim Schulte and Brian McGuinness. New York: Barnes and Noble, 1979.

———. *Notebooks 1914–16*. Edited by Georg H. von Wright and G. E. M. Anscombe. Translated by G. E. M. Anscombe. Oxford: Basil Blackwell, 1961.

———. *On Certainty*. Edited by G. E. M. Anscombe and Georg H. von Wright. Translated by Denis Paul and G. E. M. Anscombe. New York: J. and J. Evanston, 1969.

———. *Philosophical Investigations*. 4th ed. Edited by Peter M. Hacker and Joachim Schulte. Translated by G. E. M. Anscombe et al. Oxford: Blackwell, 2009.

———. *Philosophical Remarks*. Edited by Rush Rhees. Translated by Raymond Hargreaves and Roger White. Oxford: Blackwell, 1975.

———. *Philosophy of Psychology—A Fragment* [previously known as 'Pt. II' of *Philosophical Investigations*, 4th ed]. Edited by Peter M. Hacker and Joachim Schulte. Translated by G. E. M. Anscombe, et al. Oxford: Blackwell, 2009.

———. "Remarks on Frazer's Golden Bough." In *Philosophical Occasions*, edited by James C. Klagge and Alfred Nordmann, translated by Rush Rhees, 115–55. Indianapolis: Hackett, 1993.

———. *Tractatus Logico-Philosophicus*. Translated by Charles K. Ogden. Mineola, NY: Dover, 1999.

Wundt, Wilhelm. *Lectures on Human and Animal Psychology*. Translated by James Creighton & Edward Titchener. London: George Allen, 1912.

Yannaras, Christos. *Relational Ontology*. Translated by Norman Russell. Brookline, MA: Holy Cross Orthodox Press, 2011.

Yudkowsky, Eliezer. "What Do We Mean by 'Rationality?'" *LessWrong* (blog), March 16, 2009. https://www.lesswrong.com/posts/RcZCwxFiZzE6X7nsv/what-do-we-mean-by-rationality-1.

Index